Math Contests for Grades 4, 5, and 6

Volume 7

School Years
2011-2012 through 2015-2016

Written by

Steven R. Conrad • Daniel Flegler • Adam Raichel

Published by MATH LEAGUE PRESS
Printed in the United States of America

Cover art by Bob DeRosa

First Printing, 2016

Math League Press
P.O. Box 17
Tenafly, NJ 07670-0017

ISBN 978-0-940805-21-7

Preface

Math Contests—Grades 4, 5, and 6, Volume 7 is the seventh volume in our series of problem books for grades 4, 5, and 6. The first six volumes contain the contests given in the school years 1979-1980 through 2010-2011. This volume contains contests given from 2011-2012 through 2015-2016. (You can use the order form on page 154 to order any of our 21 books.)

This book is divided into three sections for ease of use by students and teachers. You'll find the contests in the first section. Each contest consists of 30 or 35 multiple-choice questions that you can do in 30 minutes. On each 3-page contest, the questions on the 1st page are generally straightforward, those on the 2nd page are moderate in difficulty, and those on the 3rd page are more difficult. In the second section of the book, you'll find detailed solutions to all the contest questions. In the third and final section of the book are the letter answers to each contest. In this section, you'll also find rating scales you can use to rate your performance.

Many people prefer to consult the answer section rather than the solution section when first reviewing a contest. We believe that reworking a problem when you know the answer (but *not* the solution) often leads to increased understanding of problem-solving techniques.

Each year we sponsor an Annual 4th Grade Mathematics Contest, an Annual 5th Grade Mathematics Contest, and an Annual 6th Grade Mathematics Contest. A student may participate in the contest on grade level or for any higher grade level. For example, students in grades 4 and 5 (or below) may participate in the 6th Grade Contest. Starting with the 1991-92 school year, students have been permitted to use calculators on any of our contests.

Steven R. Conrad, Daniel Flegler, & Adam Raichel, contest authors

Acknowledgments

For her continued patience and understanding, special thanks to Marina Conrad, whose only mathematical skill, an important one, is the ability to count the ways.

For demonstrating the meaning of selflessness on a daily basis, special thanks to Grace Flegler.

To Jeannine Kolbush, who did an awesome proofreading job, thanks!

Table Of Contents

The Contests

•••••••••••••••••••••••

2011-2012 through 2015-2016

4th Grade Contests

2011-2012 through 2015-2016

2011-2012 Annual 4th Grade Contest

Spring, 2012

4

Instructions

- **Time** Do *not* open this booklet until you are told by your teacher to begin. You will have only *30 minutes* working time for this contest. You might be *unable* to finish all 30 questions in the time allowed.
- **Scores** Please remember that *this is a contest, and not a test*—there is no "passing" or "failing" score. Few students score as high as 24 points (80% correct). Students with half that, 12 points, *should be commended!*
- **Format and Point Value** This is a multiple-choice contest. Each answer is an A, B, C, or D. Write each answer in the *Answer Column* to the right of each question. A correct answer is worth 1 point. Unanswered questions receive no credit. You **may** use a calculator.

 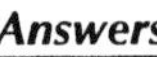

1. Which of the following is *not* a whole number?

   A) $20 + 12$ B) $20 - 12$ C) 20×12 D) $20 \div 12$

2. Alan delivers mail to even-numbered houses only. He could deliver mail to a house whose number is

   A) 1 B) 4 C) 7 D) 11

3. $(2 + 4 + 6 + 8 + 10) - (1 + 3 + 5 + 7 + 9) =$

   A) 1 B) 5 C) 10 D) 15

4. What number is twenty-one less than forty-two?

   A) 20 B) 21 C) 22 D) 23

5. I own 2 dozen pairs of white socks. How many white socks do I own?

   A) 12 B) 24 C) 36 D) 48

6. $(7 + 6 + 5) \times (4 - 3) \times 2 =$

   A) 36 B) 72 C) 138 D) 252

7. Half of 80 divided by twice 2 is

   A) 10 B) 20 C) 40 D) 80

8. I spent half of my \$5 on 25¢ stickers and the other half on 50¢ stickers. How many stickers did I buy?

   A) 5 B) 10 C) 15 D) 20

9. At the bank, Clyde was asked to find the total amount of money in 4 bags. The 4 bags held \$987, \$625, \$777, and \$553. If Clyde rounded the amount in each bag to the nearest \$100 and then added, what was Clyde's sum?

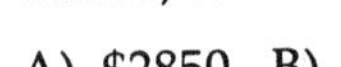

   A) \$2850 B) \$2900 C) \$2950 D) \$3000

10. The remainder of 12 345 divided by 5, plus the remainder of 54 321 divided by 10, is

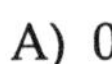

   A) 0 B) 1 C) 6 D) 11

11. How many of the numbers 1, 2, 3, 4, 5, 6, 7, 8, and 9 can be written as a sum of exactly two of the other numbers?

   A) 5 B) 6 C) 7 D) 8

Go on to the next page ⟹ **4**

 Answers

12. It's 8:15 A.M. and Larry the Bird has been listening to music for 85 minutes. At what time did he start listening?

A) 6:50 A.M. B) 7:10 A.M.
C) 7:30 A.M. D) 7:50 A.M.

12.

13. If 2 watermelons can serve 15 people, I need _?_ watermelons for 60 people.

A) 4 B) 6 C) 8 D) 12

13.

14. I have 5 nickels, 10 dimes, and 25 quarters. I have

A) \$3.00 B) \$4.25 C) \$6.00 D) \$7.50

14.

15. (The number of sides a square has) + (the number of sides a rectangle has) + (the number of sides a rhombus has) =

A) 11 B) 12 C) 13 D) 14

15.

16. Which of the following can be written as the product of an even number and an odd number?

A) 24 B) 33 C) 47 D) 95

16.

17. How many divisors of 120 are odd?

A) 2 B) 3 C) 4 D) 5

17.

18. Dylan earned \$1 on his first day selling lemonade. He earned \$2 on his second day selling lemonade. Each day after that his earnings doubled. On what day did he first earn more than \$50 on a single day?

A) day 5 B) day 6 C) day 7 D) day 8

18.

19. Chef Shemp uses 4 times as many carrots in his pot of soup as he uses in 1 salad. Three of chef's salads together have 4 fewer carrots than his pot of soup. Chef's salads have _?_ carrots each.

A) 4 B) 6 C) 8 D) 16

19.

20. The sum of the first 4 prime numbers is

A) 11 B) 16 C) 17 D) 31

20.

21. Today is a Tuesday. One hundred days from now will be a

A) Tuesday B) Thursday C) Saturday D) Sunday

21.

Go on to the next page ⟫➡ 4

22. When Mr. Smyth dances, for every 1 person in the audience applauding, there are 8 people who aren't. There are 10 people applauding, so there must be _?_ people in the audience. 22.

A) 90 B) 88 C) 80 D) 18

23. Twelve years after next year, I will be twice as old as I am now. How old am I now? 23.

A) 16 B) 15 C) 14 D) 13

24. Someone put three dimes into my pile of quarters. If I add up the value of these coins, including the dimes, the sum could be 24.

A) \$6.25 B) \$7.75 C) \$8.05 D) \$9.50

25. The radius of my circle is twice as long as one side of my square. If my square has a perimeter of 20, the diameter of my circle is 25.

A) 5 B) 10 C) 16 D) 20

26. My sister runs 10 km per hour, and I run 2 km in 15 minutes. If we both run for 2 hours, my sister will run _?_ km farther than I will. 26.

A) 2 B) 4 C) 6 D) 8

27. The largest possible sum of 4 *unequal* even numbers, none greater than 100, is 27.

A) 380 B) 388 C) 390 D) 394

28. A square with an area of 64 can be cut up into _?_ squares with a perimeter of 4. 28.

A) 8 B) 16 C) 32 D) 64

29. Each time Wanda waved her wand, 4 more stars appeared on her dress (which started with no stars). After several waves, Wanda multiplied the total number of stars then on her dress by the number of times she had waved her wand. This product *cannot* be 29.

A) 144 B) 256 C) 364 D) 676

30. If I multiply 333 333 333 333 333 by 777 777 777 777 777 and add the first and last digits of the product, the sum is 30.

A) 3 B) 5 C) 7 D) 11

The end of the contest 4

Visit our Web site at http://www.mathleague.com

Solutions on Page 73 • Answers on Page 138

2012-2013 Annual 4th Grade Contest

Spring, 2013

4

Instructions

- **Time** Do *not* open this booklet until you are told by your teacher to begin. You will have only *30 minutes* working time for this contest. You might be *unable* to finish all 30 questions in the time allowed.
- **Scores** Please remember that *this is a contest, and not a test*—there is no "passing" or "failing" score. Few students score as high as 24 points (80% correct). Students with half that, 12 points, *should be commended!*
- **Format and Point Value** This is a multiple-choice contest. Each answer is an A, B, C, or D. Write each answer in the *Answer Column* to the right of each question. A correct answer is worth 1 point. Unanswered questions receive no credit. You **may** use a calculator.

1. $2 \times 0 \times 1 \times 3 =$

   A) 0 B) 6 C) 12 D) 2013

   1.

2. Rollo delivers 3 packages to each of the 4 houses on Sixth Street. Rollo delivers a total of _?_ packages.

   A) 7 B) 12 C) 13 D) 72

   2.

3. What is the remainder when 16 + 16 + 16 + 16 is divided by 4?

   A) 16 B) 4 C) 2 D) 0

   3.

4. Which of the following is a factor of 380?

   A) 3 B) 6 C) 8 D) 10

   4.

5. If there are 8 pencils in each box, how many pencils are in 80 boxes?

   A) 10 B) 88 C) 640 D) 808

   5.

6. $(60 \div 5) \times 4 =$

   A) 3 B) 16 C) 48 D) 96

   6.

7. Stan earns two dimes for every glass of lemonade he sells. If Stan earned $20, how many glasses of lemonade did he sell?

   A) 10 B) 20 C) 40 D) 100

   7.

8. How many whole numbers are greater than 9 and less than 60?

   A) 49 B) 50 C) 51 D) 59

   8.

9. Sue sits on a seesaw waiting for her friend Seth. Seth left on a Saturday and will be back seventeen days later. Seth will be back on a

   A) Sunday B) Tuesday
   C) Thursday D) Friday

   9.

10. The greatest odd factor of 30 is

    A) 5 B) 6 C) 15 D) 21

    10.

11. Wayne goes to bed exactly 65 minutes after 8:30 P.M. At what time does Wayne go to bed?

    A) 9:05 P.M. B) 9:25 P.M. C) 9:35 P.M. D) 9:45 P.M.

    11.

Go on to the next page ⟫➡ **4**

12. Roy has rowed his rowboat 1000 m from where he started. Roy has rowed his rowboat _?_ cm. — 12.

A) 10 B) 100
C) 10000 D) 100000

13. My pocketful of coins includes quarters, dimes, nickels, and exactly 8 pennies. Of the following, which could be the total value of my pocketful of coins? — 13.

A) $14.56 B) $16.32 C) $18.85 D) $21.93

14. $4 \times 4 \times 20 \times 20 = 80 \times$ _?_ — 14.

A) 80 B) 20 C) 4 D) 2

15. If the sum of the lengths of the sides of a rhombus is 24, then each side of the rhombus has a length of — 15.

A) 3 B) 4 C) 6 D) 8

16. If 20 years ago Allen was half as old as he is today, how old was he 10 years ago? — 16.

A) 20 B) 30 C) 40 D) 50

17. If the sum of 7 whole numbers is even, at most _?_ of the numbers could be odd. — 17.

A) 6 B) 4 C) 3 D) 1

18. (10 hundreds) + (10 ones) = _?_ tens — 18.

A) 10 B) 101 C) 110 D) 1010

19. Sam loves spaghetti and meatballs. He prepares a plate of spaghetti with some meatballs. If the number of meatballs is divisible by 4, 5, 6, 7, and 8, there must be at least _?_ meatballs. — 19.

A) 210 B) 420 C) 840 D) 6720

20. The number that is 25 less than the number that is 50 less than 125 is — 20.

A) 0 B) 25 C) 50 D) 75

21. The product of 2 odd numbers is always — 21.

A) divisible by 3 B) odd C) prime D) even

Go on to the next page ⟶ 4

22. Charlie grills 3 hot dogs for every 8 hamburgers he grills. If he grills 48 hamburgers, he grills _?_ hot dogs.

A) 18 B) 43 C) 80 D) 128

22.

23. Today is my birthday. If my age in months is 99 greater than my age in years, how many years old am I now?

A) 9 B) 11 C) 12 D) 14

23.

24. The radius of a circle is half the length of the side of a square. The square's perimeter is equal to the diameter of the circle multiplied by

A) 2 B) 4 C) 8 D) 16

24.

25. When each of the following is divided by 8, only _?_ has a remainder that is a prime number.

A) 548 B) 569 C) 678 D) 778

25.

26. My aunt can fold 16 paper cranes in 4 minutes. My uncle can fold 15 paper cranes in 5 minutes. How long would it take them to fold 42 cranes if they work together at those rates?

A) 6 minutes B) 9 minutes C) 12 minutes D) 13 minutes

26.

27. If $1 + 3 + 5 + 7 + 9 + \ldots + 99 = 2500$, then $3 + 5 + 7 + 9 + \ldots + 101 =$

A) 2500 B) 2600 C) 2601 D) 2700

27.

28. Alfonse's high chair is 10 times as tall as his cat. His cat is 8 times as tall as his pet rat. His rat is 6 times as tall as his pet cricket. If his cricket is 4 mm tall, how tall is Alfonse's high chair?

A) 28 mm B) 480 mm C) 960 mm D) 1920 mm

28.

29. Ray runs every other day. If he ran for the first time last month on a Monday, then he ran for the tenth time last month on a

A) Monday B) Tuesday C) Friday D) Sunday

29.

30. How many of the whole numbers less than 100 are 10 greater than an odd whole number?

A) 45 B) 46 C) 90 D) 91

30.

The end of the contest 4

Visit our Web site at http://www.mathleague.com

Solutions on Page 77 • Answers on Page 139

Math League Press, P.O. Box 17, Tenafly, New Jersey 07670-0017

2013-2014 Annual 4th Grade Contest

4

Spring, 2014

Instructions

- **Time** You will have only *30 minutes* working time for this contest. You might be *unable* to finish all 30 questions in the time allowed.
- **Scores** Please remember that *this is a contest, and not a test*—there is no "passing" or "failing" score. Few students score as high as 24 points (80% correct). Students with half that, 12 points, *deserve commendation!*
- **Format and Point Value** This is a multiple-choice contest. Each answer is an A, B, C, or D. Write each answer in the *Answer Column* to the right of each question. A correct answer is worth 1 point. Unanswered questions receive no credit. You **may** use a calculator.

2013-2014 4TH GRADE CONTEST

Answers

1. $2 \times 0 \times 1 \times 4 =$

   A) 0 B) 7 C) 8 D) 2014

2. Cubby just graduated, but his claws make it hard for him to hold his diploma! He has five claws on each of his four paws, for a total of _?_ claws.

   A) 9 B) 16 C) 18 D) 20

3. If 2 years ago I was 3 years old, how old am I now?

   A) 4 B) 5 C) 6 D) 23

4. I had $1, and then I spent 55¢. How much do I have left?

   A) 45¢ B) 55¢ C) 65¢ D) 75¢

5. $8 + (60 \div 4) =$

   A) 15 B) 17 C) 22 D) 23

6. A summer vacation exactly 91 days long lasts for how many weeks?

   A) 9 B) 11 C) 12 D) 13

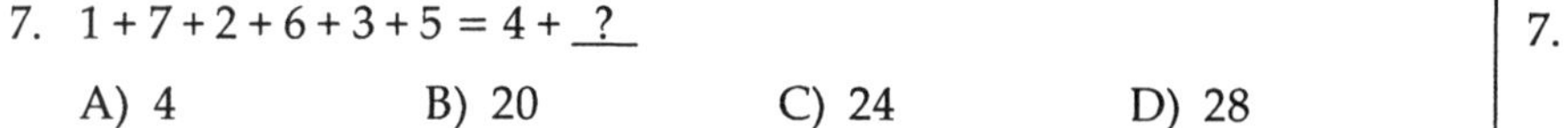

7. $1 + 7 + 2 + 6 + 3 + 5 = 4 +$ _?_

   A) 4 B) 20 C) 24 D) 28

8. What is 3 less than 50 more than 400?

   A) 87 B) 91 C) 447 D) 474

9. How many prime numbers are less than 10?

   A) 2 B) 3 C) 4 D) 5

10. Five nickels and ten dimes have the same total value as _?_ quarters.

    A) 5 B) 10 C) 15 D) 25

11. Caleb the dog dreams he has 12 dozen bones. If he buries them in pairs, one pair per hole, he will have to dig _?_ holes.

    A) 24 B) 72 C) 144 D) 288

12. There are 42 424 fans in the stands at a football game. Rounded to the nearest 100, that is _?_ fans.

    A) 42 000 B) 42 100 C) 42 400 D) 42 500

Go on to the next page ⟹ **4**

13. Cary the crooning crow started singing his songs at 9:45 PM. If he finished at 11:10 PM that night, he was singing for __?__ minutes.

    A) 65 B) 75 C) 85 D) 95

14. The remainder when 124 is divided by 8 is

    A) 2 B) 4 C) 5 D) 6

15. The hundreds digit of the product 2014×400 is

    A) 0 B) 5 C) 6 D) 8

16. I am going to travel 200 km at a rate of 60 km per hour. How many minutes will my trip take?

    A) 60 B) 200 C) 260 D) 320

17. Greta was 110 cm tall 2 years ago, when she was 10 cm taller than her brother. They both have grown, but now Greta is 10 cm shorter than her brother. If her brother grew 40 cm in those two years, Greta is now __?__ cm tall.

    A) 120 B) 130 C) 140 D) 150

18. $1000 \div 100 = 10 \times$ __?__

    A) 1 B) 10 C) 100 D) 1000

19. Starting on January 1st, Sam put $2 into a jar on every day with an odd-numbered date. How much will be in the jar after she puts in her $2 on February 21st?

    A) $48 B) $50 C) $52 D) $54

20. How many whole numbers between 5 and 15 are divisible by 4?

    A) 1 B) 2 C) 3 D) 4

21. $81 + 72 + 63 + 54 = 9 \times$ __?__

    A) 20 B) 24 C) 30 D) 36

22. Chester chewed up 2 pencils for every 3 he put in his pencil holder. If there are 30 pencils in his holder, how many pencils did Chester chew up?

    A) 12 B) 15 C) 20 D) 45

Go on to the next page ⟹ **4**

23. Manny the mammoth is friends with Murray the mastodon. Manny weighs three times as much as Murray. If Manny weighs 8000 kg more than Murray, than Manny weighs _?_ kg. — 23.

A) 4000 B) 6000
C) 8000 D) 12 000

24. The sum of the three greatest whole-number divisors of 36 is — 24.

A) 66 B) 39 C) 36 D) 19

25. I have twice as many shirts as hats, and four times as many hats as scarves. If I have 24 shirts, how many scarves do I have? — 25.

A) 2 B) 3 C) 6 D) 12

26. If half the area of a square is 32, then half the perimeter of the square is — 26.

A) 16 B) 20 C) 32 D) 64

27. The coins in my pocket have a total value of $6.20. I have 5 more pennies than nickels, and equal numbers of nickels, dimes, and quarters. If I have no other coins, how many pennies do I have? — 27.

A) 10 B) 15 C) 20 D) 25

28. Professor Drake has a student write the number 2014 more than 600 times. The 2014th digit the student writes will be — 28.

A) 2 B) 0 C) 1 D) 4

29. The distance from my house to the park is 6 km, and the distance from my school to the park is 8 km. The distance from my house to my school *cannot* be _?_ km. — 29.

A) 14 km B) 10 km C) 8 km D) 1 km

30. The houses on Long Street have addresses from a low of 300 to a high of 356, including every whole number in between. The sum of all the even addresses minus the sum of all the odd addresses is — 30.

A) 28 B) 300 C) 328 D) 356

The end of the contest 4

Visit our Web site at http://www.mathleague.com

Solutions on Page 81 • Answers on Page 140

Math League Press, P.O. Box 17, Tenafly, New Jersey 07670-0017

2014-2015 Annual 4th Grade Contest

Spring, 2015

4

Instructions

- **Time** You will have only *30 minutes* working time for this contest. You might be *unable* to finish all 30 questions in the time allowed.
- **Scores** Please remember that *this is a contest, and not a test*—there is no "passing" or "failing" score. Few students score as high as 24 points (80% correct). Students with half that, 12 points, *deserve commendation!*
- **Format and Point Value** This is a multiple-choice contest. Each answer is an A, B, C, or D. Write each answer in the *Answer Column* to the right of each question. A correct answer is worth 1 point. Unanswered questions receive no credit. You **may** use a calculator.

1. Two times fourteen = four times ___?___ | 1.

A) 7 B) 10 C) 20 D) 28

2. Each nilbog from outer space has 2 hands and 2 feet, but has only 3 toes on each foot and 3 fingers on each hand. If there are 3 nilbogs, they have a combined total of ___?___ fingers and toes. | 2.

A) 9 B) 12 C) 18 D) 36

3. $(10 + 20) \times (30 + 40) =$ | 3.

A) 100 B) 650 C) 940 D) 2100

4. The number that is 34 more than 56 is | 4.

A) 90 B) 80 C) 32 D) 22

5. $7 \times 8 \times 9 \times 10 \times 11 = 8 \times 9 \times 10 \times 11 \times$ ___?___ | 5.

A) 5 B) 7 C) 12 D) 18

6. Rounded to the nearest 10, the value of 9876 is | 6.

A) 9800 B) 9870 C) 9880 D) 9900

7. What is the perimeter of the rectangle at right? | 7.

4 9

A) 13 B) 26 C) 36 D) 49

8. 5 nickels + ___?___ dimes = 5 quarters | 8.

A) 5 B) 10 C) 15 D) 20

9. The sum of 123 and 345 is | 9.

A) even B) odd C) prime D) over 500

10. Of the following, which has the greatest number of positive factors? | 10.

A) 10 B) 15 C) 20 D) 25

11. Quincy Magoo Jr. led a hike. The map said to walk 14×32 steps and stop at the cliff. Quincy walked 12×34 steps. He walked ___?___ fewer steps than the map said to. | 11.

A) 4 B) 12 C) 16 D) 40

12. Ten days after Tuesday is | 12.

A) Sunday B) Monday C) Friday D) Saturday

Go on to the next page ⟫⟫⟫➡ 4

Answers

13. Father Time is jumping up and down to celebrate a new year. On jump 1 he jumps 10 cm, on jump 2 he jumps 20 cm, and on each jump after that he goes twice as high as he did on the jump before. What number jump will be his first jump of over 1 m? 13.

A) 4 B) 5 C) 10 D) 11

14. $(88 + 88 + 88) \div 8 = 99 \div$ _?_ 14.

A) 3 B) 9 C) 11 D) 33

15. The ones digit of $1234 \times 2345 + 345 \times 456 + 56 \times 67 + 7 \times 8$ is 15.

A) 0 B) 2 C) 6 D) 8

16. I will be 8 years older than I was 3 years ago in _?_ years. 16.

A) 5 B) 8 C) 11 D) 24

17. Triple the number you get by tripling a whole number, then subtract 3. The result is always divisible by 17.

A) 9 B) 6 C) 3 D) 2

18. If I buy 6 pens for \$1.05 each and you buy 8 pencils for 90¢ each, I spend _?_ less than you do. 18.

A) \$0.90 B) \$1.00 C) \$1.10 D) \$1.20

19. Which of the following lists puts three shapes in order from most number of sides to least number of sides? 19.

A) triangle, square, pentagon B) square, triangle, rectangle
C) pentagon, rectangle, triangle D) pentagon, triangle, square

20. Tom's job is to paint walls. Earlier today he painted 7 walls. Now he is painting another wall. Later he will paint 11 more. If these are the only walls he paints today, he will have painted _?_ walls in all. 20.

A) 18 B) 19 C) 20 D) 21

21. $(4 \times 444) + (4 \times 444) = 4 \times$ _?_ 21.

A) 444 B) 666 C) 888 D) 1616

22. How many fingers are there on a dozen pairs of gloves if each glove has 5 fingers? 22.

A) 24 B) 30 C) 60 D) 120

Go on to the next page ➠ 4

23. What an exciting game! At the end of the 1st half, the Roaring Hamsters were 1 point behind the Screaming Gerbils. During the second half the Gerbils didn't score, and the Hamsters finished ahead by 7. If the Hamsters doubled their score during the second half, the Hamsters' final score was

A) 16 B) 14 C) 12 D) 8

23.

24. Jon tried to find the sum 1 + 2 + 3 + 4 + 5 + 6 + 7 + 8 + 9 + 10, but left out one number. If his sum was 48, which number did Jon leave out?

A) 6 B) 7 C) 8 D) 9

24.

25. Writing the number ten million requires __?__ more zeros than writing the number one hundred thousand.

A) 1 B) 2 C) 3 D) 4

25.

26. What is the remainder when 987 654 321 is divided by 100?

A) 21 B) 42 C) 65 D) 98

26.

27. My old printer printed 13 pages in 3 seconds. My new printer prints 21 pages in 2 seconds. In 6 seconds my new printer prints __?__ more pages than my old one could have.

A) 8 B) 16 C) 27 D) 37

27.

28. I multiply a whole number by itself, then multiply that product by itself. The ones digit of my final product *cannot* be

A) 1 B) 4 C) 5 D) 6

28.

29. Huey jumps a positive whole number of cm, and that number is divisible by both 9 and 10. The least possible value of that number has exactly __?__ whole number divisors.

A) 6 B) 8 C) 10 D) 12

29.

30. The difference between my mother's age and my father's age is 5. The sum of their ages could be

A) 42 B) 56 C) 63 D) 78

30.

The end of the contest **4**

Visit our Web site at http://www.mathleague.com

Solutions on Page 85 • Answers on Page 141

FOURTH GRADE MATHEMATICS CONTEST

Math League Press, P.O. Box 17, Tenafly, New Jersey 07670-0017

2015-2016 Annual 4th Grade Contest

Spring, 2016

4

Instructions

- **Time** You will have only *30 minutes* working time for this contest. You might be *unable* to finish all 30 questions in the time allowed.
- **Scores** Please remember that *this is a contest, and not a test*—there is no "passing" or "failing" score. Few students score as high as 24 points (80% correct). Students with half that, 12 points, *deserve commendation!*
- **Format and Point Value** This is a multiple-choice contest. Each answer is an A, B, C, or D. Write each answer in the *Answer Column* to the right of each question. A correct answer is worth 1 point. Unanswered questions receive no credit. You **may** use a calculator.

1. $4 \times 9 =$ — 1.

   A) 16×2 B) 12×3 C) 7×5 D) 38×1

2. If each fork costs $6, each spoon costs $7, and each knife costs $8, what is the total cost of 3 forks, 4 spoons, and 5 knives? — 2.

   A) $40 B) $46 C) $80 D) $86

3. When Flash the tortoise divides the number of peas he ate this morning by 5, the remainder *cannot* be — 3.

   A) 0 B) 1 C) 3 D) 5

4. The length and width of a rectangle are 20 and 16, respectively. Its perimeter is — 4.

   A) 320 B) 160 C) 72 D) 36

5. $10000 \div 200 \times \underline{\ ?\ } = 10\,000$ — 5.

   A) 100 B) 200 C) 1000 D) 2000

6. Which pair of numbers has a common factor greater than 1? — 6.

   A) 4 and 9 B) 6 and 27 C) 27 and 50 D) 33 and 100

7. 5 m + 5 cm + 5 mm = __?__ mm — 7.

   A) 5055 B) 5505 C) 5550 D) 55550

8. 33 hours and 36 minutes = __?__ minutes — 8.

   A) 1996 B) 2006 C) 2016 D) 2026

9. The product of any two multiples of 3 must be a multiple of — 9.

   A) 6 B) 9 C) 12 D) 15

10. Which of the following has an odd number of whole-number factors? — 10.

    A) 47 B) 48 C) 49 D) 50

11. Joe started with 30 books. He sold 1 book, bought 2 books, sold 3 books, bought 4 books, sold 5 books, and finally bought 6 books. After all that, how many books does he have? — 11.

    A) 9 B) 27 C) 33 D) 51

Go on to the next page ⟹ **4**

	Answers
12. On Day 1, there are 256 apples on a tree. On each day after Day 1 until one apple is left, the number of apples on the tree is half the number there were the day before. On Day _?_, there will be only one apple on the tree.  A) 10 B) 9 C) 8 D) 7	12.
13. A square whose side-length is _?_ has the same numerical value for its perimeter and its area. A) 1 B) 2 C) 4 D) 16	13.
14. What is the ones digit of $2015^{2015} + 2016^{2016}$? A) 1 B) 3 C) 5 D) 7	14.
15. $24 \times 26 \times 28 \times 30 \times 32 = 48 \times 52 \times 56 \times 60 \times$ _?_ A) 2 B) 4 C) 16 D) 64	15.
16. What time is 11999 hours after 3 P.M.? A) 2 A.M. B) 2 P.M. C) 4 A.M. D) 4 P.M.	16.
17. Which of the following is a factor of both 12345678 and 123456789? A) 2 B) 7 C) 9 D) 11	17.
18. The product of any whole number and 2 is always A) prime B) composite C) odd D) even	18.
19. If my age 15 years ago was 15, my age 15 years from now will be A) 15 B) 30 C) 45 D) 60	19.
20. What a busy week! Each of my 4 English assignments took me 45 minutes to complete, and each of my 6 math assignments took 40 minutes. On just these, I spent a total of _?_ hours on homework assignments. A) 6 B) 7 C) 420 D) 430	20.
21. $2016 \times 20 + 2016 \times 40 + 2016 \times 60 = 2016 \times$ _?_ A) 120 B) 240 C) 2016 D) 48000	21.
22. How many thousands of seconds are there in 365 days? A) 31536 B) 525600 C) 1892160 D) 3536000	22.

Go on to the next page ⟫⟫➜ 4

23. Tom and Jerry had a fishing contest. By noon Tom had caught 47 more fish than Jerry, so Tom went to sleep and caught no more fish. Jerry kept fishing, and the total number of fish he caught all day was 19 more than Tom's total. If the total number of fish Jerry caught that day was 3 times as many as he had caught by noon, Jerry caught a total of _?_ fish. — 23.

A) 33 B) 80 C) 99 D) 102

24. The sum of five consecutive whole numbers is 280. What is the sum of the next five consecutive whole numbers? — 24.

A) 285 B) 305 C) 405 D) 425

25. The product of one million and 1000 million has _?_ zeros at the end. — 25.

A) 15 B) 16 C) 17 D) 54

26. _?_ is a factor of $1 \times 2 \times 3 \times 4 \times 5 \times 6 \times 7 \times 8 \times 9 \times 10$. — 26.

A) 71 B) 73 C) 75 D) 77

27. For every 7 soccer balls Elena bought for the gym, she bought 4 basketballs. If she bought 35 soccer balls, she bought a **total** of _?_ balls. — 27.

A) 20 B) 35 C) 45 D) 55

28. How many whole numbers between 1 and 100 are 3 times a prime? — 28.

A) 9 B) 10 C) 11 D) 12

29. Barb likes to help her father. She dusts every 3 days, sweeps every 4 days, and cooks dinner every 5 days. If she does all 3 chores on a Sunday, she next does all 3 on the same day on a — 29.

A) Wednesday B) Thursday
C) Friday D) Saturday

30. How many different primes are in the prime factorization of 2016? — 30.

A) 3 B) 4 C) 7 D) 8

The end of the contest ✍ 4

Visit our Web site at http://www.mathleague.com

Solutions on Page 89 • Answers on Page 142

5th Grade Contests

2011-2012 through 2015-2016

Math League Press, P.O. Box 17, Tenafly, New Jersey 07670-0017

2011-2012 Annual 5th Grade Contest

Spring, 2012

5

Instructions

- **Time** Do *not* open this booklet until you are told by your teacher to begin. You will have only *30 minutes* working time for this contest. You might be *unable* to finish all 30 questions in the time allowed.
- **Scores** Please remember that *this is a contest, and not a test*—there is no "passing" or "failing" score. Few students score as high as 24 points (80% correct). Students with half that, 12 points, *should be commended!*
- **Format and Point Value** This is a multiple-choice contest. Each answer is an A, B, C, or D. Write each answer in the *Answer Column* to the

	Answers
1. Fran's forklift lifts 2012 kg at a time. In 4 lifts, Fran's forklift will lift _?_ kg. A) 503 B) 2016 C) 4024 D) 8048	1.
2. $(22 + 44 + 66) \div (2 + 4 + 6) =$ A) 10 B) 11 C) 33 D) 44	2.
3. Each of the following results in an even number *except* A) $952 + 136$ B) $952 - 136$ C) $952 \div 136$ D) 952×136	3.
4. My pet frog jumps 3 m per jump. If it wants to jump from one end of a 100 m field to another, the least number of jumps it will take is A) 33 B) 34 C) 70 D) 97	4.
5. 100 hundreds + 10 tens + 1 one = A) 111 B) 1101 C) 1011 D) 10101	5.
6. If I have 2500 quarters, then I have A) \$100 B) \$500 C) \$625 D) \$1000	6.
7. Which date is 100 days after November 6th? A) February 14 B) February 15 C) February 16 D) February 17	7.
8. What is the remainder when $12 + 34 + 56 + 89 + 90$ is divided by 10? A) 0 B) 1 C) 4 D) 9	8.
9. Connie counts from 1 to 20. What is the sum of the prime numbers she counts? A) 29 B) 30 C) 77 D) 78	9.
10. Flash the photographer's lucky number is 8. Today is lucky because as of today he has now taken $(888\,888 + 888) \div 8$ photos. He has taken _?_ photos. A) 111111 B) 111222 C) 888999 D) 111111111	10.
11. Ron's work day is half over at 1:15 P.M. If he starts work at 9:30 A.M., his work day ends at A) 4:30 P.M. B) 5:00 P.M. C) 5:30 P.M. D) 9:30 P.M.	11.

Go on to the next page ⟫⟫➔ 5

Answers

12. Inspector Ivan is looking for the largest number that is a factor of both 36 and 90. He is looking for

A) 2 B) 3 C) 9 D) 18

12.

13. I just drove 60 km in 20 minutes. I drove at an average rate of _?_ km per hr.

A) 20 B) 30 C) 120 D) 180

13.

14. What is the ones digit of the product $80 \times 70 \times 60 \times 50 \times 40 \times 30 \times 20 \times 10 \times 5 \times 2$?

A) 0 B) 2 C) 5 D) 7

14.

15. Starting with a square whose perimeter is 36, I form a rectangle by doubling the length of two of the sides and tripling the length of the other two sides. This rectangle's perimeter is

A) 60 B) 90 C) 216 D) 486

15.

16. Bram bought 60 beans for $3.00. At this price, 100 beans cost

A) $3.50 B) $4.00 C) $5.00 D) $5.50

16.

17. My age is 3 years less than 5 times my sister's age. If I am 27, the sum of my age and my sister's age is

A) 33 B) 36 C) 39 D) 42

17.

18. I ate lunch every day of my 12-week summer break. I ate lunch outdoors only on every Saturday, Sunday, and Tuesday. I ate lunch indoors on _?_ days.

A) 24 B) 36 C) 48 D) 60

18.

19. The average of 4000 fours is four times the average of 2000 _?_ .

A) ones B) twos C) fours D) eights

19.

20. When I add together the number of sides of a quadrilateral, a trapezoid, and a parallelogram, I get a total of

A) 12 B) 13 C) 14 D) 15

20.

21. Wally had guests 4 days in a row. He had 26 guests on the 1st day, 32 on the 2nd day, and the same number on the 3rd day as on the 4th day. If he averaged 30 guests per day on those 4 days, how many guests did he have on the 3rd day?

A) 29 B) 30 C) 31 D) 32

21.

Go on to the next page ⟹ 5

Answers

22. Emily is 5 years younger than Gene, and Gene is 5 years younger than his car. If the ages of Emily, Gene, and the car add up to 165, the car is _?_ years old.

A) 50 B) 55 C) 60 D) 65

22.

23. What is the least possible remainder when an even number is divided by 7?

A) 0 B) 1 C) 5 D) 6

23.

24. $4 : 14 = 14 :$ _?_

A) 4 B) 24 C) 49 D) 114

24.

25. There are four books on my shelf. The almanac and the dictionary together weigh 4310 g. The biography and the cookbook together weigh 2325 g. If the almanac and the biography together weigh 2795 g, what is the weight of the cookbook and the dictionary together?

A) 2620 g B) 3270 g C) 3840 g D) 4100 g

25.

26. $100 - 98 + 96 - 94 + 92 - 90 + \ldots + 8 - 6 + 4 - 2 + 0 =$

A) 26 B) 50 C) 52 D) 100

26.

27. An average of 40 campers sign up for each of the 6 sports at camp. If each camper signs up for exactly 3 sports, how many campers are there?

A) 80 B) 120 C) 160 D) 180

27.

28. I am in line with 24 people behind me, including my friend. She has 24 people in front of her, including me. If 10 people are in line between my friend and me, all together how many people are in line?

A) 37 B) 38 C) 56 D) 58

28.

29. Baby Carl's dad is making enough food to feed 12 babies for 21 days! At that rate, the same amount of food could feed 9 babies for _?_ days.

A) 16 B) 18 C) 24 D) 28

29.

30. The sum of 2 positive whole numbers is 3 times their difference. When the larger is divided by the smaller, the quotient is

A) 2 B) 3 C) 4 D) 5

30.

The end of the contest **5**

Visit our Web site at http://www.mathleague.com

Solutions on Page 95 • Answers on Page 143

Math League Press, P.O. Box 17, Tenafly, New Jersey 07670-0017

2012-2013 Annual 5th Grade Contest

Spring, 2013

5

Instructions

- **Time** Do *not* open this booklet until you are told by your teacher to begin. You will have only *30 minutes* working time for this contest. You might be *unable* to finish all 30 questions in the time allowed.
- **Scores** Please remember that *this is a contest, and not a test*—there is no "passing" or "failing" score. Few students score as high as 24 points (80% correct). Students with half that, 12 points, *should be commended!*
- **Format and Point Value** This is a multiple-choice contest. Each answer is an A, B, C, or D. Write each answer in the *Answer Column* to the right of each question. A correct answer is worth 1 point. Unanswered questions receive no credit. You **may** use a calculator.

Answers

1. Sue hasn't struck out since 18 days before Saturday. That day was a

   A) Tuesday B) Wednesday
   C) Thursday D) Friday

2. $(1 + 2 + 3) \times 10 = 30 + 20 +$ _?_

   A) 10 B) 11 C) 33 D) 44

3. I listened to 6 songs before the one I'm listening to now, and I will listen to 6 more after this one. All together, that's _?_ songs.

   A) 11 B) 12 C) 13 D) 14

4. 100 hundreds ÷ 10 tens =

   A) 10 B) 100 C) 1000 D) 10000

5. $9 + 99 + 999 = 9 \times$ _?_

   A) 111 B) 112 C) 122 D) 123

6. I created 30 characters, 3 for each video game I own. That means I own _?_ video games.

   A) 10 B) 33 C) 40 D) 90

7. If I add the number of sides that a hexagon has to the number of sides that a _?_ has, then the sum is odd.

   A) rhombus B) square C) pentagon D) quadrilateral

8. $40 + 30 \times 20 + 10 \times 0 =$

   A) 0 B) 150 C) 640 D) 1400

9. My older brother is 6 years older than I am, and the sum of our ages is 30. How old is my older brother?

   A) 12 B) 15 C) 18 D) 21

10. Don paid for 5 tropical punches with a $50 bill and got $16 in change. He paid _?_ per tropical punch.

    A) $5.20 B) $6.80
    C) $8.20 D) $8.80

11. The average of one dozen and two dozen is

    A) 13 B) 18 C) 24 D) 36

Answers
1.
2.
3.
4.
5.
6.
7.
8.
9.
10.
11.

Go on to the next page ⟹ 5

12. At normal speed, it takes Manuel exactly one hour and 46 minutes to play a trombone concerto. Playing at twice that speed, it would take Manuel _?_ minutes to play the concerto.

A) 53 B) 73 C) 83 D) 212

12.

13. If I triple _?_ and then subtract 60, I get 180.

A) 40 B) 60 C) 70 D) 80

13.

14. There are a total of 2013 students enrolled at 8 high schools. If there are 234 students at each of 4 of the schools, then there are a total of _?_ students at the other 4 schools.

A) 1077 B) 1123 C) 1234 D) 1443

14.

15. Three different books are arranged in a line on my bookshelf. In how many different orders can these books be arranged?

A) 3 B) 4 C) 5 D) 6

15.

16. A square piece of paper has a perimeter of 36 cm. What is the area of a square piece of paper with twice that perimeter?

A) 72 cm^2 B) 108 cm^2 C) 144 cm^2 D) 324 cm^2

16.

17. I have equal numbers of quarters, dimes, and nickels. These coins could have a total value of any of the following EXCEPT

A) \$2.40 B) \$3.80 C) \$4.40 D) \$5.20

17.

18. Of the following, _?_ has the greatest number of whole number factors.

A) 6 B) 9 C) 12 D) 16

18.

19. The least common multiple of 10 and 24 plus the greatest common factor of 10 and 24 equals

A) 121 B) 122 C) 241 D) 242

19.

20. There are 5 cars for every 3 trucks parked in a lot. If there is a total of 120 cars and trucks parked in the lot, there are _?_ cars there.

A) 24 B) 45 C) 75 D) 80

20.

21. Sven is skiing at a rate of 600 m/min. That equals a rate of _?_ cm/sec.

A) 100 B) 600
C) 1000 D) 60000

21.

22. Maria had 28 dreams last month. If 16 of them involved monkeys, 15 involved squirrels, and 4 involved no animals, then at least how many dreams involved both monkeys and squirrels?

A) 3 B) 7 C) 9 D) 11

23. The lengths of three consecutive sides of a _?_ could be 3, 3, and 8.

A) triangle B) square
C) parallelogram D) trapezoid

24. I have 500 pennies. If I spend 6 pennies a day until I can no longer do so, at the end of one of the days I will have exactly _?_ pennies left.

A) 6 B) 8 C) 10 D) 12

25. A "combo" ticket to enter the fair and ride unlimited rides is \$30. A "per ride" ticket costs \$12.50 to enter and \$5 per ride. For a "combo" ticket to cost less than a "per ride" ticket, a person must go on at least _?_ rides.

A) 3 B) 4 C) 6 D) 7

26. The ones digit of $9 \times 8 \times 7 \times 6 \times 5 \times 4 \times 3 \times 3 \times 4 \times 5 \times 6 \times 7 \times 8 \times 9$ is

A) 0 B) 4 C) 6 D) 9

27. A football team scores an average of 31 points per game in its first four games and an average of 30 points per game in its first five games. How many points did the team score in its fifth game?

A) 26 B) 27 C) 28 D) 29

28. The 7 people in my mailbox leave. I write X for each man and O for each woman as they leave. I have 3 X's and 4 O's, with no 2 X's in a row. There are _?_ different orders in which the X's and O's could be written.

A) 4 B) 6 C) 8 D) 10

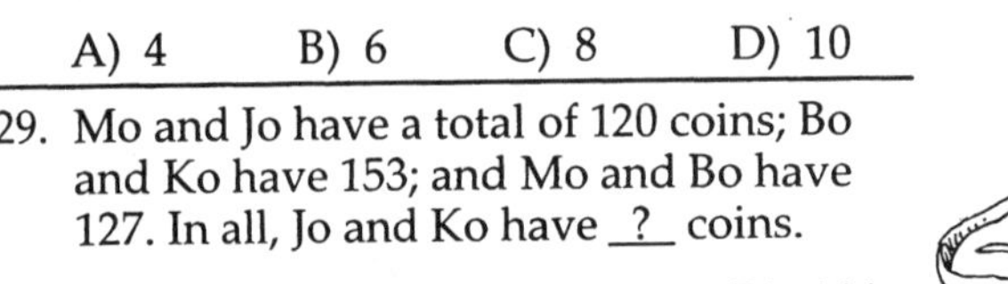

29. Mo and Jo have a total of 120 coins; Bo and Ko have 153; and Mo and Bo have 127. In all, Jo and Ko have _?_ coins.

A) 106 B) 128 C) 135 D) 146

30. The largest perimeter a rectangle made of 100 2 x 2 squares can have is

A) 88 B) 100 C) 400 D) 404

The end of the contest **5**

Visit our Web site at http://www.mathleague.com
Solutions on Page 99 • Answers on Page 144

Math League Press, P.O. Box 17, Tenafly, New Jersey 07670-0017

2013-2014 Annual 5th Grade Contest

Spring, 2014

5

Instructions

- **Time** Do *not* open this booklet until you are told by your teacher to begin. You will have only *30 minutes* working time for this contest. You might be *unable* to finish all 30 questions in the time allowed.
- **Scores** Please remember that *this is a contest, and not a test*—there is no "passing" or "failing" score. Few students score as high as 24 points (80% correct). Students with half that, 12 points, *should be commended!*
- **Format and Point Value** This is a multiple-choice contest. Each answer is an A, B, C, or D. Write each answer in the *Answer Column* to the right of each question. A correct answer is worth 1 point. Unanswered questions receive no credit. You **may** use a calculator.

1. Larry's bird wished him a happy graduation by calling in his ear, singing 4 notes each time it called. The bird called 8 times, waited, and then called 4 more times. It sang a total of __?__ notes.

   A) 3 B) 12 C) 16 D) 48

   1.

2. $2014 - 1014 + 3014 - 2014 =$

   A) 0 B) 1000 C) 2000 D) 2014

   2.

3. The sum of the measures in degrees of two of the angles in a rectangle is

   A) 90 B) 180 C) 200 D) 360

   3.

4. $10 + 9 \times 8 - 7 \times 6 =$

   A) 40 B) 110 C) 114 D) 870

   4.

5. Of the following numbers, which is closest to 2014?

   A) 1914 B) 1983 C) 2056 D) 3014

   5.

6. The largest prime factor of 72 is

   A) 3 B) 7 C) 36 D) 72

   6.

7. Sam drives three times as fast as Dave. If Dave drives 200 km in 6 hours, Sam drives 200 km in __?__ hours.

   A) 2 B) 4 C) 12 D) 18

   7.

8. $6 \times 4 =$ __?__ $\div 4$

   A) 6 B) 12 C) 24 D) 96

   8.

9. Paolo's grandma walks him to school. She drops him off at 8:18 AM and picks him up at 2:36 PM. Paolo spends __?__ minutes at school.

   A) 258 B) 378 C) 418 D) 618

   9.

10. If 6 cans of Sweet Stuff Soda all together contain 96 teaspoons of sugar, then there are a total of __?__ teaspoons of sugar in 15 cans.

    A) 192 B) 208 C) 240 D) 288

    10.

11. $88 \times 88 = 8 \times$ __?__

    A) 11×2 B) $11 + 11$ C) 11×11 D) $11 \times 11 \times 8$

    11.

12. The largest possible sum of two different two-digit numbers is

    A) 21 B) 99 C) 197 D) 198

    12.

Go on to the next page ⟫➡ **5**

13. On the first 6 days of a certain week, Alexander's ragtime band played 2, 16, 3, 15, 4, and 14 songs. The band also played on the 7th day of that week and averaged 9 songs per day for the entire week. How many songs did the band play on the 7th day of the week? 13.

A) 5 B) 7 C) 9 D) 11

14. When the number 789 678 567 456 is added to the number 987 876 765 654, how many digits does the sum have? 14.

A) 12 B) 13 C) 24 D) 25

15. Four friends buy four freakishly large orders of French fries at a fast food franchise. When they pay with a $20 bill, they get $4.44 in change. How much did each order of fries cost? 15.

A) $1.39 B) $3.89 C) $3.92 D) $7.78

16. All of the people in my town carpool to a big party. There are 5 people in each car except for the last one, in which there are only 2 people. There could be ___?___ people in my town. 16.

A) 4351 B) 5215 C) 5616 D) 6462

17. The ones digit of $106 \times 107 \times 108 \times 109 \times 110$ is 17.

A) 0 B) 2 C) 4 D) 8

18. On a school field trip, there are 2 adults for every 9 children. If there are 99 people on the trip, ___?___ of them must be adults. 18.

A) 9 B) 11 C) 18 D) 22

19. (# of sides a rhombus has) × (# of sides a square has) = (# of sides a rectangle has) × (# of sides a ___?___ has) 19.

A) triangle B) trapezoid C) hexagon D) octagon

20. Ann numbered paper hearts from 1 to 100. Her boyfriend Wilson likes only odd numbers with odd tens digits. Ann sent him all the hearts with numbers he likes (and no others), so Ann sent him ___?___ hearts. 20.

A) 23 B) 25 C) 30 D) 45

21. How many non-zero digits are in the product of 25 000 000 and 40 000 000? 21.

A) 4 B) 3 C) 2 D) 1

Go on to the next page »»» 5

22. Rich eats his favorite sandwich only on the first day of each month. Eight days after eating one, he calculates how long he will have to wait for his next. He must wait at least __?__ more days.

A) 19 B) 20 C) 22 D) 23

22.

23. If you triple my age and add three, then triple that sum, the final product is 99. How old was I exactly three years ago?

A) 7 B) 10 C) 13 D) 21

23.

24. Which of the following numbers has a prime number of whole-number factors?

A) 6 B) 12 C) 36 D) 49

24.

25. Nine chairs are in a straight line and numbered 1 to 9 from left to right. Five girls and four boys sit in the chairs so that no girl is next to another girl. A boy could be sitting in the chair with which number?

A) 1 B) 3 C) 4 D) 5

25.

26. Dividing a certain two-digit number by 10 leaves a remainder of 9, and dividing it by 9 leaves a remainder of 8. The sum of the number's two digits is

A) 7 B) 9 C) 13 D) 17

26.

27. A rectangle with a perimeter of 24 can be divided into 3 identical squares. What is the area of the rectangle?

A) 18 B) 24 C) 27 D) 48

27.

28. How many whole numbers less than 1000 can be written as the product of 3 consecutive whole numbers?

A) 10 B) 11 C) 15 D) 21

28.

29. Bill the Borzoi holds up a blank 10 by 12 canvas. The greatest number of 4 by 6 rectangles that can be drawn on Bill's canvas without overlap is

A) 2 B) 3 C) 4 D) 5

29.

30. How many whole numbers between 5000 and 6000 consist of four different digits that decrease from left to right?

A) 3 B) 10 C) 69 D) 120

30.

The end of the contest **5**

Math League Press, P.O. Box 17, Tenafly, New Jersey 07670-0017

2014-2015 Annual 5th Grade Contest

Spring, 2015

5

Instructions

- **Time** Do *not* open this booklet until you are told by your teacher to begin. You will have only *30 minutes* working time for this contest. You might be *unable* to finish all 30 questions in the time allowed.
- **Scores** Please remember that *this is a contest, and not a test*—there is no "passing" or "failing" score. Few students score as high as 24 points (80% correct). Students with half that, 12 points, *should be commended!*
- **Format and Point Value** This is a multiple-choice contest. Each answer is an A, B, C, or D. Write each answer in the *Answer Column* to the right of each question. A correct answer is worth 1 point. Unanswered questions receive no credit. You **may** use a calculator.

Answers

1. Mary Hartman has been shouting for 3 hours. She has been shouting for ___?___ minutes.

   A) 90 B) 150 C) 180 D) 300

2. $6 + (6 \div 6) + (6 \times 6) - 6 =$

   A) 37 B) 42 C) 43 D) 48

3. When 2015 is divided by 2014, the remainder is

   A) 0 B) 1 C) 4 D) 2013

4. What is the perimeter of a hexagon if each side has a length of 6?

   A) 12 B) 24 C) 30 D) 36

5. $12 \times 12 = 36 \times$ ___?___

   A) 4 B) 6 C) 9 D) 36

6. At Up 'n Down Burgers, a hamburger costs $5.50 and a soda costs $0.75. The total cost of 3 hamburgers and 2 sodas is

   A) $17.00 B) $17.50 C) $18.00 D) $18.50

7. How many whole numbers are greater than 20 but less than 50?

   A) 28 B) 29 C) 30 D) 31

8. $11 + 22 + 33 + 44 + 55 + 66 - 11 \times 11 = 11 \times$ ___?___

   A) 6 B) 7 C) 10 D) 20

9. If today is Tuesday, then 20 days from today will be

   A) Sunday B) Monday C) Tuesday D) Wednesday

10. If I have 4 ten-dollar bills and 1 twenty-dollar bill, my bills are worth an average of ___?___ per bill.

    A) $12 B) $14 C) $15 D) $30

11. Brian is shocked to find out that his neighbor's collection of Hawaiian shirts is 3 times as big as his collection! If his neighbor has 48 more Hawaiian shirts than Brian, his neighbor has ___?___ Hawaiian shirts.

    A) 24 B) 60 C) 64 D) 72

12. The number of hours in 10 complete days equals the number of minutes in ___?___ hours.

    A) 4 B) 12 C) 24 D) 30

Go on to the next page ⟹ 5

13. Rocky and Apollo were the first two babies born at Balboa General Hospital this year. Rocky was first born, and 50 minutes later Apollo was born at 1:35 P.M. At what time was Rocky born?

A) 12:45 P.M. B) 12:55 P.M.
C) 1:25 P.M. D) 2:25 P.M.

13.

14. If I start with 876 and round to the nearest 10, then multiply by 2, then round to the nearest 100, I end up with

A) 1700 B) 1740 C) 1800 D) 1860

14.

15. How many multiples of 9 are between 10 and 100?

A) 9 B) 10 C) 11 D) 12

15.

16. The sum of the ten-thousands digit and the hundreds digit of 123 456 is

A) 5 B) 6 C) 7 D) 8

16.

17. Yesterday half of the students in my class wore jeans, twelve wore dresses, and the remaining eighth of the students in my class wore khaki pants. How many students are in my class?

A) 24 B) 26 C) 28 D) 32

`17.

18. My family can build 3 snowmen in 45 minutes. How long would it take us to build 7 snowmen?

A) 1 hr, 5 min B) 1 hr, 15 min C) 1 hr, 25 min D) 1 hr, 45 min

18.

19. $1 \times 100 \times 3 \times 100 \times 5 \times 100 = 100 \times \underline{\ ?\ }$

A) 3×5 B) 8×100 C) 200×15 D) 300×500

19.

20. Albert played in a very long concert, and each song after the first had 10 notes more than the song before. If the first three songs that Albert played had a total of 2025 notes, the next three songs that he played had a total of _?_ notes.

A) 2055 B) 2085 C) 2115 D) 3025

20.

21. What is the difference between the least common multiple of 4 and 14 and the greatest common factor of 4 and 14?

A) 26 B) 18 C) 10 D) 4

21.

Go on to the next page ⟫⟫➡ 5

22. Helios is from Greece, and to honor its national colors he sells only blue and white balloons. If yesterday he sold 5 blue balloons for every 3 white balloons he sold, he might have sold exactly _?_ balloons in all.

A) 33 B) 45 C) 56 D) 71

22.

23. In Langlandia, each citizen speaks exactly 2 of the 6 official languages. If there are only 3000 citizens of this small country, an average of how many people speak each language?

A) 900 B) 1000 C) 1800 D) 2100

23.

24. Three identical squares, each with area 4, can be laid side by side to form a rectangle with a width of 2 and a perimeter of

A) 10 B) 12 C) 14 D) 16

24.

25. There are exactly 3 prime numbers between

A) 10 and 20 B) 20 and 30 C) 30 and 40 D) 40 and 50

25.

26. If the perimeter of a triangle is 28, which of the following *cannot* be the lengths of two of its sides?

A) 7 and 8 B) 8 and 9 C) 7 and 14 D) 11 and 11

26.

27. Dawn has been asked by her teacher to write the number that is the product of $600 \times 500 \times 400 \times 300$. She must write _?_ non-zero digits.

A) 1 B) 2 C) 3 D) 4

27.

28. Winkle is a very old tortoise! At his birthday today, his age in months and his age in years add up to 2015. How many years old is Winkle?

A) 155 B) 160 C) 165 D) 167

28.

29. How many three-digit whole numbers have a ones digit equal to the sum of the hundreds digit and the tens digit?

A) 10 B) 25 C) 30 D) 45

29.

30. It takes 144 workers 60 hours to paint a bridge. Working at the same rate, how many hours would 108 workers require to do the job?

A) 45 B) 65 C) 80 D) 96

30.

The end of the contest ✍ **5**

Visit our Web site at http://www.mathleague.com
Solutions on Page 107 • Answers on Page 146

Math League Press, P.O. Box 17, Tenafly, New Jersey 07670-0017

2015-2016 Annual 5th Grade Contest

Spring, 2016

5

Instructions

- **Time** Do *not* open this booklet until you are told by your teacher to begin. You will have only *30 minutes* working time for this contest. You might be *unable* to finish all 30 questions in the time allowed.
- **Scores** Please remember that *this is a contest, and not a test*—there is no "passing" or "failing" score. Few students score as high as 24 points (80% correct). Students with half that, 12 points, *should be commended!*
- **Format and Point Value** This is a multiple-choice contest. Each answer is an A, B, C, or D. Write each answer in the *Answer Column* to the right of each question. A correct answer is worth 1 point. Unanswered questions receive no credit. You **may** use a calculator.

1. Which of the following is greatest?

   A) $20 + 16$ B) $20 - 16$ C) 20×16 D) $20 \div 16$

   1.

2. If I eat 3 meals each day, then I eat _?_ meals in two weeks.

   A) 6 B) 14 C) 21 D) 42

   2.

3. $(111 \div 3) + (111 \div 3) + (111 \div 3) =$

   A) 37 B) 74 C) 111 D) 333

   3.

4. Of the following, _?_ is *not* a whole number.

   A) 0.3×5 B) 0.4×5
   C) 0.5×4 D) 3×5

   4.

5. If two angles in a right triangle have equal measures, each measures

   A) 30° B) 45° C) 60° D) 75°

   5.

6. Algebra books cost $12.50 each, and geometry books cost $14 each. How much do 6 algebra books and 5 geometry books cost in total?

   A) $145.00 B) $146.50 C) $150.00 D) $151.50

   6.

7. How many whole numbers are greater than 30 and less than 70?

   A) 38 B) 39 C) 40 D) 41

   7.

8. $25 \times 1 + 25 \times 3 + 25 \times 5 + 25 \times 7 + 25 \times 9 = 25 \times$ _?_

   A) 24 B) 25 C) 125 D) 945

   8.

9. $1 + 2 + 3 + 4 + 5 + 995 + 996 + 997 + 998 + 999 =$

   A) 5000 B) 5001 C) 5002 D) 5003

   9.

10. Last season, Larry scored 3 times as many points as Mo. Together they scored a total of 100 points. How many points did Larry score?

    A) 25 B) 50 C) 75 D) 100

    10.

11. In one week I spent a total of $120. The next week, I spent a total of $132. I spent an average of _?_ dollars per **day** during the two weeks.

    A) 18 B) 36 C) 126 D) 252

    11.

12. Which of the following is a prime number?

    A) 2013 B) 2015 C) 2016 D) 2017

    12.

13. Jack and Jill blew up a lot of balloons! For every 3 balloons Jack blew up, Jill blew up 5. If Jack blew up 21 balloons, then Jill blew up _?_ balloons.

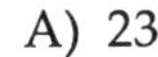
A) 23 B) 25 C) 35 D) 56

13.

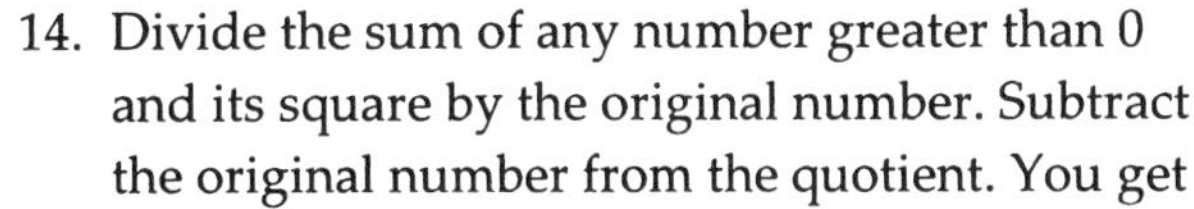
14. Divide the sum of any number greater than 0 and its square by the original number. Subtract the original number from the quotient. You get

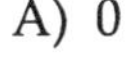
A) 0 B) 1 C) 2 D) 3

14.

15. How many whole numbers less than 20 are divisible by 6 but not by 3?

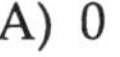
A) 0 B) 1 C) 2 D) 3

15.

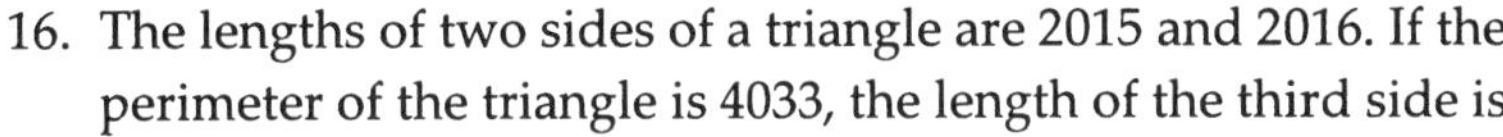
16. The lengths of two sides of a triangle are 2015 and 2016. If the perimeter of the triangle is 4033, the length of the third side is

A) 2 B) 2017 C) 2018 D) 2019

16.

17. Tom the Cat catches a mouse every 12 240 seconds. Today he caught his first mouse at 5:00 A.M. He will catch his third mouse at

A) 11:24 A.M. B) 11:48 A.M. C) 12:40 P.M. D) 3:12 P.M.

17.

18. $3 \times 6 \times 9 \times 12 = 1 \times 2 \times 3 \times 4 \times$ _?_

A) 3 B) 9 C) 27 D) 81

18.

19. If a whole number is divisible by 111, then it must be divisible by

A) 5 B) 7 C) 11 D) 37

19.

20. The sides of a right triangle have lengths of 9, 40, and 41. What is the area of the triangle?

A) 180 B) 184.5 C) 360 D) 820

20.

21. Yesterday was Tim's crazy shopping day! For every 3 red shirts he bought, he bought 4 yellow shirts and 5 green shirts. He might have bought a total of _?_ red, green, and yellow shirts yesterday.

A) 30 B) 40 C) 50 D) 120

21.

22. Doubling the radius of a circle causes the area of the circle to be multiplied by

A) 1 B) 2 C) 4 D) 8

22.

23. If Ms. Jackson writes the 9-letter phrase "MATHISFUN" on the blackboard ten times, what will be the 50th letter she writes? — 23.

A) H B) I C) N D) U

24. If the length of a side of a square is 1 m, the area of this square is _?_ cm^2. — 24.

A) 10^2 B) 10^4 C) 10^6 D) 10^9

25. Jin loves carrots! Yesterday she ate 1/2 of her carrots, and today she ate 2/3 of the remaining carrots. She then discovered that she had 12 carrots left. Yesterday she must have started with _?_ carrots. — 25.

A) 36 B) 48 C) 60 D) 72

26. Di's vehicle collection consists of unicycles, bicycles, and tricycles. Together, the vehicles have 2016 wheels. If she has 333 unicycles and 666 bicycles, how many tricycles does she have? — 26.

A) 117 B) 339 C) 351 D) 999

27. If the sum of 2016 different whole numbers is odd, at most how many of the numbers can be even? — 27.

A) 0 B) 1 C) 2014 D) 2015

28. Which of the following is a factor of $1 \times 2 \times 3 \times \ldots \times 35 \times 36$? — 28.

A) 37 B) 53 C) 62 D) 82

29. How many whole numbers between 100 and 999 have 8 as the product of their digits? — 29.

A) 8 B) 9 C) 10 D) 11

30. Clark C. Rook is the accountant for the zoo. He knows that the sum of the two whole numbers that he was supposed to write down is 50, but he can't remember what the numbers are! At least he wrote down the product of the two numbers. Of the following, which could be the product of the two numbers? — 30.

A) 478 B) 479 C) 480 D) 481

The end of the contest 5

Visit our Web site at http://www.mathleague.com

Solutions on Page 111 • Answers on Page 147

6th Grade Contests

2011-2012 through 2015-2016

Math League Press, P.O. Box 17, Tenafly, New Jersey 07670-0017

2011-2012 Annual 6th Grade Contest

Tuesday, February 21 or 28, 2012

6

Instructions

- **Time** Do *not* open this booklet until you are told by your teacher to begin. You might be *unable* to finish all 35 questions in the 30 minutes allowed.
- **Scores** Please remember that *this is a contest, and not a test*—there is no "passing" or "failing" score. Few students score as high as 28 points (80% correct). Students with half that, 14 points, *should be commended!*
- **Format, Point Value, & Eligibility** Every answer is an A, B, C, or D. Write answers in the *Answers* column. A correct answer is worth 1 point. Unanswered questions get no credit. You **may** use a calculator.

 Answers

1. Terry the tuba player just played an odd number of notes. He could have played _?_ notes.

   A) 2012 B) 2211 C) 3456 D) 4664

2. The average of _?_ and 8 is 10.

   A) 6 B) 9 C) 12 D) 18

3. The tens digit of 345 plus the hundreds digit of 456 equals

   A) 8 B) 9 C) 10 D) 11

4. My first day of vacation is May 10. My last day of vacation is May 20 of the same year. How many days of vacation do I have?

   A) 9 B) 10 C) 11 D) 12

5. How many minutes are in two full days?

   A) 1440 B) 1660 C) 2000 D) 2880

6. What is the least common multiple of 12 and 45?

   A) 57 B) 90 C) 180 D) 540

7. Dina pays 50 cents per dozen erasers. Buying 72 erasers costs her

   A) $1.50 B) $3.00 C) $30.00 D) $36.00

8. The sum of an even prime number and an odd prime number is 9. The odd prime number must be

   A) 1 B) 5 C) 6 D) 7

9. The sum of the largest and smallest whole-number divisors of 36 is

   A) 12 B) 15 C) 20 D) 37

10. There are 400 drivers stuck in a traffic jam. If 160 of them are late for work, what percent of the drivers are late for work?

    A) 20% B) 40% C) 60% D) 80%

11. The number of multiples of 4 between 1 and 222 is _?_ more than the number of multiples of 5 between 1 and 222.

    A) 11 B) 22 C) 33 D) 44

12. What time is 1000 minutes after 10 A.M.?

    A) 4:40 P.M. B) 10:00 P.M.
    C) 2:40 A.M. D) 10:00 A.M.

13. Gil flies at 800 km/hour and has 4400 km to travel. It will take him _?_ hours to complete his trip.

    A) 4 B) 4.5 C) 5 D) 5.5

Answers
1.
2.
3.
4.
5.
6.
7.
8.
9.
10.
11.
12.
13.

Go on to the next page ⟫➔ **6**

14. Larry might be lying. He takes his real age, adds 10, divides by 2, subtracts 10, and multiplies by 2 to get 30, and he claims that's his age. His real age is

A) 50 B) 40 C) 30 D) 20

14.

15. A rectangle has a perimeter of 48 and a length that is 3 times its width. The width of the rectangle is

A) 6 B) 8 C) 12 D) 16

15.

16. My sister hates quarters! She gave me 50 quarters, so I gave her 50 pennies, 50 nickels, and 50 dimes. Now she wants the rest of her money. I owe her

A) \$4.50 B) \$5.50 C) \$6.50 D) \$7.50

16.

17. (8.462 rounded to the nearest tenth) × (7531 rounded to the nearest hundred) =

A) 63000 B) 63450 C) 63750 D) 64005

17.

18. The ratio of boys to girls in my class is 3:4. There could be _?_ students in the class.

A) 120 B) 70 C) 40 D) 30

18.

19. Four friends each have \$1. Together with a fifth friend, the five have an average of \$10 each. How much does the fifth friend have?

A) \$9 B) \$19 C) \$46 D) \$49

19.

20. The largest divisor of 180 that is the square of an integer is _?_ greater than the smallest divisor of 180 that is the square of an integer.

A) 5 B) 27 C) 32 D) 35

20.

21. Woof the dog wears number 11. When I divide my number by 11, the remainder is 3. What is the remainder when I divide four times my number by 11?

A) 1 B) 3 C) 7 D) 12

21.

22. $99 \times 88 \times 77 \times 66 \times 55 = 9 \times 8 \times 7 \times 6 \times 5 \times$ _?_

A) 11 B) 11^4 C) 11^5 D) 11^6

22.

23. The largest whole number for which $3{:}5 < 12{:}$_?_ is true is

A) 19 B) 20 C) 21 D) 24

23.

24. What is the largest of 9 consecutive whole numbers whose sum is 99?

A) 9 B) 11 C) 13 D) 15

24.

25. My house is 4 km from my sister's house, and my sister's house is 9 km from my school. The distance between my house and my school *cannot* be

A) 4 km B) 5 km C) 8 km D) 10 km

25.

Go on to the next page ⟹ **6**

Answers

26. Chemist Karl wants to combine two chemicals at a time, trying every such combination he can come up with using the 7 chemicals in his kit. In how many ways can Karl combine his chemicals?

A) 14 B) 21 C) 42 D) 49

26.

27. Each of the following numbers is a divisor of 2 468 024 680 246 802 468 024 680 *except*

A) 4 B) 6 C) 8 D) 20

27.

28. If I multiply three different prime numbers, the product must have __?__ positive divisors.

A) 3 B) 5 C) 6 D) 8

28.

29. $2^{2000} + 2^{2000} + 3^{3000} + 3^{3000} + 3^{3000} =$

A) $4^{2000} + 9^{3000}$ B) $2^{4000} + 3^{9000}$ C) $4^{2001} + 9^{3002}$ D) $2^{2001} + 3^{3001}$

29.

30. I drop a nickel when I start walking at one end of a field 100 m long, and drop another nickel every 5 m until I drop the last one at the other end of the field. What is the total value of the nickels I have dropped?

A) \$0.95 B) \$1.00 C) \$1.05 D) \$1.10

30.

31. How many of the 1000 whole numbers from 1 to 1000 are both even and the square of a whole number?

A) 15 B) 16 C) 30 D) 31

31.

32. If one-half of one-quarter of a square has an area of 8, what is the perimeter of the square?

A) 16 B) 32 C) 48 D) 64

32.

33. The sum of the first 100 positive multiples of 4 is __?__ more than the sum of the first 100 positive multiples of 3.

A) 100 B) 400 C) 1200 D) 5050

33.

34. The ratio of Og's clubs to skins is 5:3. If he trades 4 of his clubs for 2 more skins, the ratio of clubs to skins will be 8:7. How many more clubs than skins does Og have before any trading occurs?

A) 8 B) 9 C) 10 D) 12

34.

35. What is the sum of the remainders when 100, 101, 102, . . . , 998, 999, and 1000 are each divided by 9?

A) 3600 B) 3601 C) 4500 D) 4501

35.

The end of the contest  6

Visit our Web site at http://www.mathleague.com

Solutions on Page 117 • Answers on Page 148

Math League Press, P.O. Box 17, Tenafly, New Jersey 07670-0017

2012-2013 Annual 6th Grade Contest

Tuesday, February 19 or 26, 2013

6

Instructions

- **Time** Do *not* open this booklet until you are told by your teacher to begin. You might be *unable* to finish all 35 questions in the 30 minutes allowed.
- **Scores** Please remember that *this is a contest, and not a test*—there is no "passing" or "failing" score. Few students score as high as 28 points (80% correct). Students with half that, 14 points, *should be commended!*
- **Format, Point Value, & Eligibility** Every answer is an A, B, C, or D. Write answers in the *Answers* column. A correct answer is worth 1 point. Unanswered questions get no credit. You **may** use a calculator.

1. Pete the pilot flew 28 times last month. If 21 of his flights were at night, how many of his flights were not at night?

   A) 7 B) 21 C) 28 D) 49

   1.

2. The sum 12 + 34 + 56 equals each of the following *except*

   A) 46+56 B) 12+90
   C) 34+68 D) 46+68

   2.

3. If I double the number of pens in my backpack and add 5, I get 23. How many pens do I have in my backpack?

   A) 9 B) 14 C) 36 D) 56

   3.

4. $65 - (43 + 21) = (65 - 43) - \underline{\ ?\ }$

   A) 1 B) 12 C) 21 D) 34

   4.

5. The dime and quarter in my hand combined with the coins in my pocket total one dime less than $1. In my pocket is

   A) 45¢ B) 55¢ C) 65¢ D) 75¢

   5.

6. Wednesday is five days after my party. On what day is my party?

   A) Friday B) Sunday C) Monday D) Tuesday

   6.

7. Which of the following is the sum of two prime numbers?

   A) 11 B) 17 C) 23 D) 31

   7.

8. Each of my shoes weighs the same. If 2 of my shoes weigh 12 kg together, then the total weight of 12 of my shoes is

   A) 2 kg B) 24 kg C) 36 kg D) 72 kg

   8.

9. $25 \times 25 = 5 \times 5 \times \underline{\ ?\ }$

   A) 2 B) 5 C) 10 D) 25

   9.

10. (Six dozen) + (one dozen pairs) = __?__ sets of three

    A) 48 B) 32 C) 24 D) 12

    10.

11. When Giggles the Clown correctly counts the dots on his costume in groups of 4, there are 3 left over. There could be __?__ dots all together.

    A) 31 B) 32 C) 33 D) 34

    11.

12. What time is 420 minutes before 4 P.M.?

    A) 4:00 A.M. B) 7:00 A.M.
    C) 9:00 A.M. D) 11:40 A.M.

    12.

13. 10 hundreds + 10 tens + 10 ones =

    A) 111 B) 1101 C) 1110 D) 101010

    13.

Go on to the next page ⟹ 6

14. Professor Quack had 7 more students this year than he had last year. If he had a total of 43 students in both years combined, how many students did he have this year? | 14.

A) 18 B) 25 C) 32 D) 36

15. All together, 27 trapezoids have the same number of sides as _?_ triangles. | 15.

A) 16 B) 18 C) 27 D) 36

16. In my garden, I have 6 roses for every 5 daisies, and those are the only flowers I have. If I have 66 flowers, how many of them are roses? | 16.

A) 11 B) 22 C) 30 D) 36

17. The sum of two different odd numbers and an even number could be | 17.

A) 52 B) 61 C) 65 D) 77

18. On a Sunday I put two rabbits in a cage. If the number of rabbits in the cage doubled every day, on what day did the cage first have more than 100 rabbits in it? | 18.

A) Thursday B) Friday C) Saturday D) Sunday

19. A pomegranate costs 4 times as much as a pawpaw. If one pomegranate costs 50¢ more than 2 pawpaws, then the pomegranate costs | 19.

A) 50¢ B) 75¢ C) $1 D) $1.50

20. If I triple _?_ and divide the result by 6, the quotient is 18. | 20.

A) 9 B) 36 C) 72 D) 108

21. $11 + 12 + 13 + 14 + 15 + 16 = 11 + 22 + 33 + 44 + 55 + 66 -$ _?_ | 21.

A) 50 B) 100 C) 150 D) 200

22. If Bob jumps 15 additional times, the total number of his jumps will be 3 times what it was 3 jumps ago. Bob has jumped _?_ times all together. | 22.

A) 12 B) 18 C) 21 D) 24

23. The total value of 10 nickels and 9 dimes equals the total value of 5 quarters and _?_ pennies. | 23.

A) 4 B) 5 C) 14 D) 15

24. How many numbers between 1 and 100 are equal to 5 times an odd number? | 24.

A) 9 B) 10 C) 11 D) 19

25. The sum of the remainders of $123 \div 4$, $234 \div 5$, and $345 \div 2$ is | 25.

A) 3 B) 6 C) 8 D) 12

Go on to the next page 6

26. If Marlon the mailman had sunny weather on exactly 12 of 30 days last month, on what percent of days was the weather *not* sunny?

A) 36% B) 40% C) 60% D) 64%

26.

27. Last month I spent $24 on magnets that cost 80¢ each, and this month I spent $24 on magnets that cost $1.20 each. The average cost per magnet was

A) $0.92 B) $0.96 C) $1.00 D) $1.04

27.

28. On a number line, __?__ is the same distance from 1.75 as it is from 7.25.

A) 2.75 B) 3.25 C) 3.75 D) 4.5

28.

29. $2^3 \times 3^4 \times 4^5 \times 6^7 \times 9^{10} =$

A) $2^{15} \times 3^{21}$ B) $2^{20} \times 3^{31}$ C) $2^{15} \times 3^{40}$ D) $2^{105} \times 3^{280}$

29.

30. In a garage, the ratio of red cars to black cars is 8:5, and the ratio of black cars to white cars is 3:4. The minimum number of cars in the garage is

A) 20 B) 59 C) 74 D) 91

30.

31. The sum of 6 consecutive integers, the largest of which is 30, is equal to the sum of 10 consecutive integers, the largest of which is

A) 17 B) 18 C) 21 D) 26

31.

32. If a radius of a circle whose area is 36π cm^2 equals the width of a rectangle, and the diameter of the circle is half the length of the rectangle, then the perimeter of the rectangle is

A) 60 cm B) 90 cm C) 144 cm D) 172 cm

32.

33. I wrote a list of consecutive positive integers beginning with 1. I then removed all multiples of 4, and I had 2345 integers left. What was the largest integer on my list after the numbers were removed?

A) 3126 B) 3127 C) 3129 D) 3130

33.

34. At the start of my temporary job, I needed to load an average of 120 boxes a day in order to finish my job on time. At first I loaded 90 boxes a day. I then had 6 days left to load the remaining 1200 boxes. How many days did I have in all for this temporary job?

A) 10 B) 16 C) 22 D) 26

34.

35. Each day last week I counted 50% more leaves than I had counted the day before. If I counted 2430 leaves last Friday, how many had I counted the Sunday before that Friday?

A) 160 B) 240 C) 280 D) 320

35.

The end of the contest 6

Visit our Web site at http://www.mathleague.com

Solutions on Page 121 • Answers on Page 149

Math League Press, P.O. Box 17, Tenafly, New Jersey 07670-0017

2013-2014 Annual 6th Grade Contest

Tuesday, February 18 or 25, 2014

6

Instructions

- **Time** Do *not* open this booklet until you are told by your teacher to begin. You might be *unable* to finish all 35 questions in the 30 minutes allowed.
- **Scores** Please remember that *this is a contest, and not a test*—there is no "passing" or "failing" score. Few students score as high as 28 points (80% correct). Students with half that, 14 points, *should be commended!*
- **Format, Point Value, & Eligibility** Every answer is an A, B, C, or D. Write answers in the *Answers* column. A correct answer is worth 1 point. Unanswered questions get no credit. You **may** use a calculator.

1. The band's trombone plays 2013 notes, the trumpet plays 2014 notes, and the tuba plays 218 notes. That's a total of _?_ notes.
   A) 6245 B) 6045 C) 4245 D) 645

2. $(50-20)+(70-40)+(90-60) = 3\times$ _?_
   A) 10 B) 30 C) 90 D) 270

3. The remainder when (999 999 999 + 666 666 + 333 + 1) is divided by 3 is
   A) 0 B) 1 C) 2 D) 3

4. All together, one dollar and one quarter and one dime and one nickel have the same value as _?_ nickels.
   A) 140 B) 70 C) 35 D) 28

5. $20-5\times2=2\times$ _?_
   A) 5 B) 15 C) 25 D) 30

6. My vacation will start three weeks and two days after yesterday. If today is Tuesday, my vacation will start on what day?
   A) Monday B) Tuesday C) Wednesday D) Thursday

7. $4\times6\times8\times10=2\times3\times4\times5\times$ _?_
   A) 2 B) 6 C) 8 D) 16

8. Which one does *not* equal 20 when rounded to the nearest 10?
   A) 18.26 B) 21.42 C) 24.68 D) 25.12

9. When I split the cost of a video game equally with 4 friends, we each pay $12.00. How much more do each of us pay if I split the cost equally with only 3 friends?
   A) $3.00 B) $4.00 C) $15.00 D) $16.00

10. The number halfway between 4 and 26 is
    A) 11 B) 13 C) 15 D) 17

11. At 5:00 P.M. on a Friday, Hal got locked in the vault. He was in there for 5040 minutes before it opened. On what day did it open?
    A) Sunday B) Monday
    C) Tuesday D) Wednesday

12. 1 000 000 =
    A) 10^5 B) 10^6 C) 10^7 D) 10^8

13. 150% of 20 = 200% of
    A) 15 B) 25 C) 30 D) 40

Answers: 1. 2. 3. 4. 5. 6. 7. 8. 9. 10. 11. 12. 13.

Go on to the next page ⟹ 6

14. The new experimental cannon fired its cannonball a total distance of 2000 mm. That is the same as a distance of _?_ cm.

A) 20000 B) 200 C) 20 D) 2

14.

15. Four 2×8 rectangles have a total area equal to that of a square of perimeter

A) 8 B) 16 C) 32 D) 64

15.

16. The sum of the prime factors of 231 is

A) 21 B) 22 C) 152 D) 383

16.

17. I see a km marker at the end of each km I drive. The markers are numbered consecutively from 1 through 100. How many markers are more than 30 km from the marker numbered 50?

A) 38 B) 39 C) 40 D) 41

17.

18. (ten-thousands digit of 123456789) × (hundreds digit of 87654321) =

A) 8 B) 10 C) 12 D) 15

18.

19. Of 60 people at a school board meeting, 24 are men. The ratio of women to men at the meeting is

A) 3:2 B) 2:3 C) 11:6 D) 6:11

19.

20. The sum of all the positive integer divisors of 18 is

A) 18 B) 20 C) 38 D) 39

20.

21. I rode my bicycle a total of 450 km in 15 hours. At this rate I would travel _?_ km in 30 minutes.

A) 15 B) 30 C) 60 D) 900

21.

22. In 8 years Mac's grandmother will be twice as old as she was 29 years ago. How old will Mac's grandmother be in 4 years?

A) 62 B) 66 C) 70 D) 74

22.

23. If the sum of 7 consecutive integers is 63, the sum of the largest and the smallest of the 7 integers is

A) 18 B) 24 C) 42 D) 64

23.

24. $\sqrt{3^2 + 4^2} =$

A) 5^2 B) 3×4 C) $3 + 4$ D) 5

24.

25. Of the first 100 positive whole numbers, the ratio of the number of multiples of 8 to the number of multiples of 4 is

A) 2:1 B) 12:25 C) 13:25 D) 1:2

25.

Go on to the next page ⟫⟫➡ 6

26. Craig the Croc cried crocodile tears at exactly 15% of the movies he saw last year, but saw 34 movies that did not make him cry. How many of the movies that he saw made him cry?

A) 6 B) 12 C) 18 D) 40

26.

27. If two consecutive whole numbers have a different number of digits, then their product must be a multiple of

A) 4 B) 11 C) 15 D) 100

27.

28. My bookcase has 5 shelves, and each shelf has the same number of books on it. If the books have a total of 18000 pages, and there are an average of 360 pages per book, how many books are on each shelf?

A) 10 B) 50 C) 55 D) 250

28.

29. $2^3 \times 2^5 \times 2^7 \times 2^{11} =$

A) 4^{13} B) 16^{26} C) 2^{1155} D) 16^{1155}

29.

30. A wall is painted with vertical stripes of red, orange, yellow, green, blue, indigo, and violet in that order, and then the pattern is repeated 50 times. How many of the first 100 stripes are orange?

A) 7 B) 12 C) 14 D) 15

30.

31. (the # of sides of a trapezoid) × (the # of sides of a rhombus) + (the # of sides of a hexagon) × (the # of sides of a triangle) =

A) 24 B) 28 C) 34 D) 66

31.

32. The hands of a circular clock form a __?__ angle at 6:15 P.M.

A) 82.5° B) 90° C) 97.5° D) 270°

32.

33. At Terrier Teahouse, each blend contains 2 different teas. There are 6 herbal teas and 4 teas that are not herbal on the menu. If at least 1 tea in a blend must be herbal, __?__ different blends are possible.

A) 24 B) 27 C) 39 D) 54

33.

34. It takes 36 workers 48 hours to paint a ship. Working at the same rate, how many hours would 24 workers need to paint the ship?

A) 32 B) 56 C) 64 D) 72

34.

35. The product of all whole numbers from 1 through 35 is divisible by each of the following *except*

A) 2^{18} B) $7^6 \times 3$ C) $3^{15} \times 5$ D) $10^8 \times 7$

35.

The end of the contest 6

Visit our Web site at http://www.mathleague.com

Solutions on Page 125 • Answers on Page 150

SIXTH GRADE MATHEMATICS CONTEST

Math League Press, P.O. Box 17, Tenafly, New Jersey 07670-0017

2014-2015 Annual 6th Grade Contest

Tuesday, February 17 or 24, 2015

6

Instructions

- **Time** Do *not* open this booklet until you are told by your teacher to begin. You might be *unable* to finish all 35 questions in the 30 minutes allowed.
- **Scores** Please remember that *this is a contest, and not a test*—there is no "passing" or "failing" score. Few students score as high as 28 points (80% correct). Students with half that, 14 points, *should be commended!*
- **Format, Point Value, & Eligibility** Every answer is an A, B, C, or D. Write answers in the *Answers* column. A correct answer is worth 1 point. Unanswered questions get no credit. You **may** use a calculator.

1. Three cases of 24 cans each is the same number of cans as twelve boxes of _?_ cans each. — 1.

A) 6 B) 15 C) 21 D) 96

2. A trapezoid has _?_ sides. — 2.

A) 3 B) 4 C) 5 D) 10

3. Abby put an X through 7 of the 28 days in this month. Abby put an X through _?_ of the days. — 3.

A) 0.07 B) 0.25 C) 0.28 D) 0.35

4. $60 \div 4 = 3 \times$ _?_ — 4.

A) 20 B) 12 C) 6 D) 5

5. $30 \times 40 \times 50 = 120 \times$ _?_ — 5.

A) 50 B) 200 C) 500 D) 600

6. If 3 out of 4 of the students in my class have brown eyes, and there are 24 students in my class, then how many have brown eyes? — 6.

A) 6 B) 8 C) 18 D) 21

7. Four identical squares, each with a perimeter of 36, are put together as shown to make one big square. What is the perimeter of the big square? — 7.

A) 36 B) 48 C) 72 D) 144

8. $80 + (160 + 240) \div 4 = 40 + 80 + (120 \div$ _?_ $)$ — 8.

A) 4 B) 2 C) 1 D) 0

9. In which of the following divisions is the remainder greatest? — 9.

A) $1111 \div 8$ B) $2222 \div 7$ C) $3333 \div 6$ D) $4444 \div 5$

10. Of the following, which is a factor of $20 \times 14 \times 20 \times 15$? — 10.

A) 13 B) 11 C) 9 D) 7

11. Thok has a simple plan. He will spend 50% of the day in the cave, 25% of the rest of the day on the hunt, and the remains of the day watching films outside. How many hours will Thok spend watching films? — 11.

A) 3 B) 6 C) 9 D) 25

12. $2 \times 3 \times 6 \times 36 \times 2 \times 3 \times 6 \times 36 =$ — 12.

A) 6^5 B) 6^6 C) 6^7 D) 6^8

13. I have 5 pennies, 4 nickels, 3 quarters, 2 half-dollars, and 1 dollar. The average value of one of my coins is — 13.

A) \$0.20 B) \$0.60 C) \$1.50 D) \$3.00

Go on to the next page ⟹ 6

Answers

14. Wyatt O'Vine's sheep weighs twice as much as Wyatt, and Wyatt weighs twice as much as his hat. If Wyatt, his sheep, and his hat have a combined weight of 210 kg, how much does Wyatt weigh?

A) 30 kg B) 35 kg C) 60 kg D) 70 kg

14.

15. The least common multiple of 24 and 30, minus the greatest common factor of 24 and 30, is

A) 114 B) 117 C) 234 D) 237

15.

16. Two consecutive whole numbers *cannot* have a sum that is

A) a prime B) a perfect square
C) a multiple of 13 D) an even number

16.

17. The greatest prime factor of the number of minutes in a week is

A) 5 B) 7 C) 12 D) 21

17.

18. $(12+34)\times(56+78) = 12\times(56+78) + \underline{\ ?\ }\times(56+78)$

A) 12 B) 34 C) 56 D) 78

18.

19. Seth eats an apple at 4 P.M. on Monday and then waits 100 hours before he eats another. He eats his next apple at _?_ on Friday.

A) noon B) 4 P.M. C) 8 P.M. D) midnight

19.

20. If 2 flocks equal 5 flecks, then 500 flocks equal _?_ flecks.

A) 200 B) 250 C) 1000 D) 1250

20.

21. $\left(\sqrt{64}+\sqrt{64}\right)^2 =$

A) 16 B) 64 C) 128 D) 256

21.

22. If one angle of a triangle is 10 degrees smaller than another and 20 degrees smaller than the third, what is the largest angle in the triangle?

A) 70° B) 75° C) 80° D) 90°

22.

23. In 4 hours Rob rode 120 km. He drove at an average rate of _?_ m/min.

A) 30 B) 105 C) 120 D) 500

23.

24. Clara the alligator has laid 65% of her eggs. If she has laid 13 eggs, how many eggs are left to be laid?

A) 6 B) 7 C) 13 D) 20

24.

25. If the sum of seven consecutive even numbers is 182, the smallest of the seven numbers is

A) 20 B) 23 C) 26 D) 32

25.

Go on to the next page ⟶ 6

26. While he was standing on his head, Flip decided to count backward from 777 by eights. Which of the following numbers did Flip count?

A) 123 B) 125 C) 127 D) 129

26.

27. The cost of five apples is the same as the cost of six pears. If one apple costs 15 cents more than one pear, then what is the total cost of 5 apples and 6 pears?

A) $3 B) $6 C) $9 D) $18

27.

28. What is the difference between 27 and the product of all its whole-number factors?

A) 2 B) 27 C) 2×27 D) 26×27

28.

29. The prime factorization of a whole number less than 100 is the product of at most _?_ primes (not necessarily different).

A) 3 B) 4 C) 5 D) 6

29.

30. A rectangle with sides of integer length is divided into a square region and a shaded rectangular region as shown. If the area of the shaded region is 8, the area of the entire figure is at most

A) 24 B) 64 C) 72 D) 81

30.

31. Gabriel wrote his name 100 times in the sand. What was the 100th letter he wrote?

A) a B) b C) r D) i

31.

32. I put $100 in a savings account. At the end of each year, I earned 10% interest on the balance of my account at that point. With no deposits or withdrawals, after 5 years I had _?_. (Round to the nearest dollar.)

A) $162 B) $161 C) $160 D) $150

32.

33. Mrs. Andrews enjoys feeding the birds her special mix of birdseed. She mixes 1 part sunflower seed with 3 parts sesame seed to make a total of 200 6-kg bags. If sunflower seed comes in 12-kg bags, how many such bags does she need?

A) 25 B) 34 C) 50 D) 67

33.

34. The product of 100 one hundreds is the same as the sum of _?_ one hundreds.

A) 100^{100} B) 100^{99} C) 100^{10} D) 100^{2}

34.

35. $5^{21} \times 4^{11} \div 2 =$

A) 10^{11} B) 10^{21} C) 10^{22} D) 10^{23}

35.

The end of the contest 6

Visit our Web site at http://www.mathleague.com

Solutions on Page 129 • Answers on Page 151

Math League Press, P.O. Box 17, Tenafly, New Jersey 07670-0017

2015-2016 Annual 6th Grade Contest

Tuesday, February 16 or 23, 2016

6

Instructions

- **Time** Do *not* open this booklet until you are told by your teacher to begin. You might be *unable* to finish all 35 questions in the 30 minutes allowed.
- **Scores** Please remember that *this is a contest, and not a test*—there is no "passing" or "failing" score. Few students score as high as 28 points (80% correct). Students with half that, 14 points, *should be commended!*
- **Format, Point Value, & Eligibility** Every answer is an A, B, C, or D. Write answers in the *Answers* column. A correct answer is worth 1 point. Unanswered questions get no credit. You **may** use a calculator.

1. $2 \times 2016 = 8 \times$? | 1.

A) 504 B) 508 C) 1008 D) 8064

2. Bert Sampson sold his 2 beavers for \$222.22 each. Mho gave him \$500 for them. How much change should Bert give Mho? | 2.

A) \$277.78 B) \$277.88
C) \$55.56 D) \$55.66

3. What is the sum of the measures of the angles in a trapezoid? | 3.

A) 90° B) 180° C) 270° D) 360°

4. $10 \times 20 \times 30 \times 40 = 24 \times$? | 4.

A) 10^3 B) 10^4 C) 10^5 D) 10^6

5. $1 + 2 + 3 + 4 + 996 + 997 + 998 + 999 =$ | 5.

A) 3998 B) 3999 C) 4000 D) 4001

6. Which of the following is *not* a factor of 2016? | 6.

A) 10 B) 9 C) 8 D) 7

7. What is the sum of the tenths and the hundredths digits in the number 12 345.6789? | 7.

A) 7 B) 11 C) 13 D) 15

8. If 1 out of 6 lightbulbs is defective and there are 2016 lightbulbs, how many of them are not defective? | 8.

A) 5 B) 336 C) 1680 D) 2016

9. The time twelve thousand and twelve hours after 7 A.M. is | 9.

A) 1 A.M. B) 1 P.M. C) 7 A.M. D) 7 P.M.

10. The product of two different primes has ? divisors. | 10.

A) 3 B) 4 C) 5 D) 6

11. While eating pancakes, Adam ate $\frac{1}{4}$ of them, Jerry ate $\frac{7}{20}$, Steve ate $\frac{3}{10}$, and Dan ate the rest. Who ate the greatest number of pancakes? | 11.

A) Adam B) Jerry C) Steve D) Dan

12. The greatest common factor is smallest for which of the following pairs of numbers? | 12.

A) 4 & 18 B) 5 & 25 C) 6 & 33 D) 8 & 35

Go on to the next page ⟹ 6

13. I donate a $100 bill, 2 $50 bills, 3 $20 bills, 4 $10 bills, and 5 $5 bills. If 5 people divide my money equally, each person receives

A) $37 B) $65 C) $70 D) $75

13.

14. Peter ran at 4m/s. Paul started at the same spot 6 sec. later and chased Peter at 7 m/s. Paul needed _?_ seconds to catch Peter.

A) 6 B) 7 C) 8 D) 14

14.

15. The product of two different nonzero integers cannot be

A) prime B) zero C) even D) composite

15.

16. What is the largest factor of $2^2 \times 3^3 \times 5^5 \times 7^7 \times 11^{11}$ less than 100?

A) 66 B) 77 C) 88 D) 99

16.

17. The number of hours in 10 days = the number of minutes in _?_ hours.

A) 2 B) 3 C) 4 D) 6

17.

18. If the last 2 digits of an integer are 84, the integer must be divisible by

A) 4 B) 8 C) 9 D) 16

18.

19. The maximum number of intersection points of 4 different circles is

A) 16 B) 12 C) 8 D) 6

19.

20. What percent of 20 is 50?

A) 40 B) 140 C) 200 D) 250

20.

21. The sum of 2016 integers is even. At most _?_ of them can be odd.

A) 2016 B) 2015 C) 1 D) 0

21.

22. The measure of one angle in a triangle is $\frac{1}{3}$ of a second angle in the triangle and $\frac{1}{6}$ of the third one. The measure of the largest angle is

A) 108° B) 72° C) 54° D) 18°

22.

23. When the square of the perimeter of a certain square is divided by the area of this square, the quotient is

A) 2 B) 4 C) 8 D) 16

23.

24. 1000 m per second = _?_ km per hour

A) 60 B) 360 C) 3600 D) 6000

24.

25. The circumference of a circle with radius π is

A) π B) 2π C) π^2 D) $2\pi^2$

25.

Go on to the next page 6

26. Lex had 144 coins left after giving one-third of his coins to Clark and one-third of his remaining coins to Lois. How many coins did he give Lois?

A) 72 B) 48 C) 36 D) 32

26.

27. The ones digit of the fourth power of an integer cannot be

A) 1 B) 3 C) 5 D) 6

27.

28. Penelope paid \$24.58 for pens and pencils. If pens cost 9¢ each and pencils cost 4¢ each, at most how many pens did Penelope purchase?

A) 273 B) 272 C) 271 D) 270

28.

29. My average score on 8 math tests is 90. If my average score on the first 5 tests was 87, what was my average score on the last 3 tests?

A) 96 B) 95 C) 94 D) 93

29.

30. The cost of my algebra book is 200% of that of my geometry book. The cost of my geometry book is $\frac{4}{3}$ that of my calculus book. If my calculus book costs \$21, how much do all 3 books cost together?

A) \$28 B) \$56 C) \$84 D) \$105

30.

31. Of the integers from 1 to 1000, how many are multiples of 3, 4, *and* 5?

A) 14 B) 15 C) 16 D) 17

31.

32. If the sum of 9 consecutive odd integers is 1935, what is the sum of the next 9 consecutive odd integers?

A) 2015 B) 2017 C) 2097 D) 2099

32.

33. $3^{336} \times 9^{336} \times 27^{336} =$

A) 3^{1008} B) 3^{1344} C) 3^{1680} D) 3^{2016}

33.

34. Increasing the radius of a circle by 50% increases its area by _?_%.

A) 50 B) 100 C) 125 D) 150

34.

35. A person's initials are two letters in a specific order. If everyone has initials, what is the fewest number of people who must attend a party to be sure that three of the people will have the same initials?

A) 1353 B) 1352 C) 677 D) 676

35.

The end of the contest 6

Visit our Web site at http://www.mathleague.com

Solutions on Page 133 • Answers on Page 152

Detailed Solutions

2011-2012 through 2015-2016

4th Grade Solutions

2011-2012 through 2015-2016

Math League Press, P.O. Box 17, Tenafly, New Jersey 07670-0017

Information & Solutions

Spring, 2012

4

Contest Information

- **Solutions** Turn the page for detailed contest solutions (written in the question boxes) and letter answers (written in the *Answer Column* to the right of each question).

- **Scores** Please remember that *this is a contest, and not a test*—there is no "passing" or "failing" score. Few students score as high as 24 points (80% correct); students with half that, 12 points, *deserve commendation!*

- **Answers and Rating Scales** Turn to page 138 for the letter answers to each question and the rating scale for this contest.

	Answers
1. Since $20 \div 12$ is 1 with remainder 8, it is not a whole number. A) $20 + 12 = 32$ B) $20 - 12 = 8$ C) $20 \times 12 = 240$ D) $20 \div 12$	1. D
2. Alan delivers mail to even-numbered houses only. Since even numbers end in 0, 2, 4, 6, or 8, only B is possible. A) 1 B) 4 C) 7 D) 11	2. B
3. $(2-1)+(4-3)+(6-5)+(8-7)+(10-9)=5$. A) 1 B) 5 C) 10 D) 15	3. B
4. The number that is twenty-one less than forty-two is $42 - 21 = 21$. A) 20 B) 21 C) 22 D) 23	4. B
5. I own 2 dozen pairs $= 2 \times 12$ pairs $= 24$ pairs $= 24 \times 2 = 48$ white socks. A) 12 B) 24 C) 36 D) 48	5. D
6. $(7+6+5) \times (4-3) \times 2 = 18 \times 1 \times 2 = 36$. A) 36 B) 72 C) 138 D) 252	6. A
7. Half of 80 is 40; 40 divided by twice 2 is $40 \div 4 = 10$. A) 10 B) 20 C) 40 D) 80	7. A
8. I spent 250¢ on 25¢ stickers and 250¢ on 50¢ stickers. Since 250¢ ÷ 25¢ = 10 and 250¢ ÷ 50¢ = 5, I bought $10 + 5 = 15$ stickers. A) 5 B) 10 C) 15 D) 20	8. C
9. Clyde rounded the amount of money in each bag to the nearest 100, then added. If the four bags held \$987, \$625, \$777, and \$553, Clyde's total was \$1000 + \$600 + \$800 + \$600 = \$3000. A) \$2850 B) \$2900 C) \$2950 D) \$3000	9. D
10. $12345 \div 5$ has a remainder of 0; $54321 \div 10$ has a remainder of 1. Their sum is $0 + 1 = 1$. A) 0 B) 1 C) 6 D) 11	10. B
11. I wrote 1, 2, 3, 4, 5, 6, 7, 8, and 9 on my page. Adding 1 to 2, 3, 4, 5, 6, 7, and 8 results in 3, 4, 5, 6, 7, 8, and 9. We cannot get a sum of 1 or 2. A) 5 B) 6 C) 7 D) 8	11. C

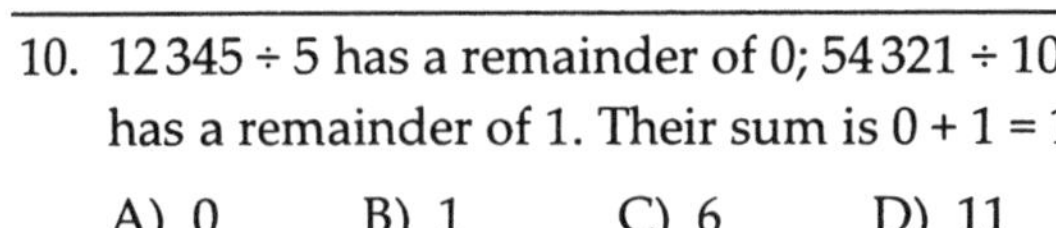

Go on to the next page 4

12. It's 8:15 A.M. He's been listening for 85 minutes = 1 hour + 25 minutes; 25 minutes before 7:15 A.M. is 6:50 A.M.

A) 6:50 A.M. B) 7:10 A.M.
C) 7:30 A.M. D) 7:50 A.M.

Answer 12. A

13. If 2 melons can serve 15 people, I need $4 \times 2 = 8$ melons for $4 \times 15 = 60$ people.

A) 4 B) 6 C) 8 D) 12

Answer 13. C

14. 5 nickels = 1 quarter; 10 dimes = 4 quarters. Total = 30 quarters = \$7.50.

A) \$3.00 B) \$4.25 C) \$6.00 D) \$7.50

Answer 14. D

15. (The number of sides a square has) + (the number of sides a rectangle has) + (the number of sides a rhombus has) = 4 + 4 + 4 = 12.

A) 11 B) 12 C) 13 D) 14

Answer 15. B

16. Since $24 = 3 \times 8$, 24 can be written as the product of an even number and an odd number. Note: Odd numbers do not have even divisors.

A) 24 B) 33 C) 47 D) 95

Answer 16. A

17. The odd divisors of 120 are 1, 3, 5, and 15. There are 4 odd divisors.

A) 2 B) 3 C) 4 D) 5

Answer 17. C

18. Dylan earned \$1 on his first day selling lemonade, then \$2 on his second day, \$4 on his third day, \$8 on his fourth day, \$16 on his fifth day, \$32 on his sixth day, and \$64 on his seventh day. The seventh day was the first day he earned more than \$50 on a single day.

A) day 5 B) day 6 C) day 7 D) day 8

Answer 18. C

19. The number of carrots in 4 salads is the same as the number of carrots in 1 pot of soup. Since 3 salads have 4 fewer carrots than 1 pot of soup, a single salad must have 4 carrots.

A) 4 B) 6 C) 8 D) 16

Answer 19. A

20. 2 + 3 + 5 + 7 = 17. Note: 1 is not prime.

A) 11 B) 16 C) 17 D) 31

Answer 20. C

21. 100 days ÷ (7 days/week) = 14 weeks and 2 days. It's 2 days after Tues.

A) Tuesday B) Thursday C) Saturday D) Sunday

Answer 21. B

Go on to the next page ⟫⟫ 4

Solution	Answers
22. When Mr. Smyth dances, for every 1 person in the audience applauding, there are 8 people who aren't. There are 10 people applauding, so 8×10 aren't. That's a total of $10 + 80 = 90$. A) 90 B) 88 C) 80 D) 18	22. A
23. In 13 years my age doubles, so it goes from 13 to 26. I am now 13. A) 16 B) 15 C) 14 D) 13	23. D
24. Subtract \$0.30 from each choice and find the one divisible by \$0.25; \$8.05 − \$0.30 = \$7.75, a multiple of \$0.25. A) \$6.25 B) \$7.75 C) \$8.05 D) \$9.50	24. C
25. If the perimeter of my square is 20, each side has length $20 \div 4 = 5$. The radius is twice as long; it's 10. The diameter is twice the radius = 20. A) 5 B) 10 C) 16 D) 20	25. D
26. I run 2 km in 15 minutes, or 8 km in 1 hour. In 2 hours, I will run 16 km and my sister will run 20 km. She will run 4 km farther than I will. A) 2 B) 4 C) 6 D) 8	26. B
27. The largest possible sum is $100 + 98 + 96 + 94 = 388$. A) 380 B) 388 C) 390 D) 394	27. B
28. A square of perimeter 4 has side-length 1. A square of area 64 can be cut into 64 squares of perimeter 4. A) 8 B) 16 C) 32 D) 64	28. D
29. The product after 1 wave = $1\times4 = 4 = (1\times2)^2$, after 2 waves = $2\times8 = 16 = (2\times2)^2$, after 3 waves = $3\times12 = 36 = (3\times2)^2$, . . . , after 6 waves = $6\times24 = 144 = (6\times2)^2$, . . . , after 8 waves = $8\times32 = 256 = (8\times2)^2$, . . . , after 13 waves = $13\times52 = 676 = (13\times2)^2$. A) 144 B) 256 C) 364 D) 676	29. C
30. Just as for 3×7, 33×77, and 333×777, the first digit is a 2, the last digit is a 1, and the sum is $2 + 1 = 3$. A) 3 B) 5 C) 7 D) 11	30. A

The end of the contest 4

Visit our Web site at http://www.mathleague.com

Math League Press, P.O. Box 17, Tenafly, New Jersey 07670-0017

Information & Solutions

Spring, 2013

4

Contest Information

- **Solutions** Turn the page for detailed contest solutions (written in the question boxes) and letter answers (written in the *Answer Column* to the right of each question).

- **Scores** Please remember that *this is a contest, and not a test*—there is no "passing" or "failing" score. Few students score as high as 24 points (80% correct); students with half that, 12 points, *deserve commendation!*

- **Answers and Rating Scales** Turn to page 139 for the letter answers to each question and the rating scale for this contest.

	Answers
1. The product of a number multiplied by 0 is 0. A) 0 B) 6 C) 12 D) 2013	1. A
2. Rollo delivers 3 packages to each of the 4 houses on Sixth Street. Rollo delivers a total of $3 \times 4 = 12$ packages. A) 7 B) 12 C) 13 D) 72	2. B
3. Since $16 \div 4$ has remainder 0, the remainder is $0+0+0+0=0$. A) 16 B) 4 C) 2 D) 0	3. D
4. Since $380 = 10 \times 38$, 10 is a factor of 380. A) 3 B) 6 C) 8 D) 10	4. D
5. There are $80 \times 8 = 640$ pencils in 80 boxes. A) 10 B) 88 C) 640 D) 808	5. C
6. $(60 \div 5) \times 4 = 12 \times 4 = 48$. A) 3 B) 16 C) 48 D) 96	6. C
7. Stan earns 20¢ for every glass of lemonade he sells. Stan earns \$20, which is 2000¢, so he sells 2000¢ ÷ 20¢ = 100 glasses of lemonade. A) 10 B) 20 C) 40 D) 100	7. D
8. There are 60 whole numbers from 0 to 59. That's 50 without 0 to 9. A) 49 B) 50 C) 51 D) 59	8. B
9. Seventeen days is the same as 14 days + 3 days. Since 14 days is two weeks, Seth will be back 3 days after Saturday. He will be back on Tuesday. A) Sunday B) Tuesday C) Thursday D) Friday	9. B
10. The smallest even factor is 2; $30 \div 2 = 15$, the greatest odd factor. A) 5 B) 6 C) 15 D) 21	10. C
11. Wayne goes to bed exactly 65 minutes after 8:30 P.M. Since 65 minutes = 1 hour + 5 minutes, Wayne will go to bed at 9:35 P.M. A) 9:05 P.M. B) 9:25 P.M. C) 9:35 P.M. D) 9:45 P.M.	11. C

Go on to the next page ⟫⟫ 4

12. Roy has rowed his rowboat 1000 m from where he started. Since 1 m = 100 cm, Roy rowed 1000 × 100 = 100000 cm.

A) 10 B) 100 C) 10000 D) 100000

12. D

13. My pocketful of coins includes quarters, dimes, nickels, and exactly 8 pennies. Since 8 pennies is 3 more than 5¢, my amount of money must end with a 3 or an 8.

A) $14.56 B) $16.32 C) $18.85 D) $21.93

13. D

14. (4 × 20) × (4 × 20) = 80 × 80.

A) 80 B) 20 C) 4 D) 2

14. A

15. If the sum of the lengths of the sides of a rhombus is 24, then each side of the rhombus has a length of 24 ÷ 4 = 6.

A) 3 B) 4 C) 6 D) 8

15. C

16. If 20 years ago Allen was half as old as he is today, then today he is 40. Thus, 10 years ago he was 30.

A) 20 B) 30 C) 40 D) 50

16. B

17. If the sum of 7 whole numbers is even, there must be an even number of odd numbers. The total number of odd numbers could be 6.

A) 6 B) 4 C) 3 D) 1

17. A

18. (10 hundreds) + (10 ones) = 1000 + 10 = 1010 = 101 tens.

A) 10 B) 101 C) 110 D) 1010

18. B

19. Sam prepares a plate of spaghetti with so many meatballs that the number of meatballs is divisible by 4, 5, 6, 7, and 8. The lcm of 4, 5, 6, 7, and 8 is 4 × 5 × 3 × 7 × 2 = 840.

A) 210 B) 420 C) 840 D) 6720

19. C

20. The number that is 50 less than 125 is 75. The number that is 25 less than 75 is 50.

A) 0 B) 25 C) 50 D) 75

20. C

21. The product of 2 odd numbers, such as 5 × 7 = 35, is always odd.

A) divisible by 3 B) odd C) prime D) even

21. B

Go on to the next page ⟫⟫➡ 4

	Answers
22. Charlie grills 3 hot dogs for every 8 hamburgers he grills. If he grills 48 hamburgers, that is 6 groups of 8 burgers. So he grills $6 \times 3 = 18$ hot dogs. A) 18 B) 43 C) 80 D) 128	22. A
23. Today is my birthday. My age in months is 12 times my age in years and is also 99 greater. Since $9 \times 12 = 108$ and $9 + 99 = 108$, I am 9 years old. A) 9 B) 11 C) 12 D) 14	23. A
24. If a radius of a circle is half the length of a side of a square, a diameter is equal to the length of one side. The perimeter is 4 times the diameter. A) 2 B) 4 C) 8 D) 16	24. B
25. The remainder upon division by 8 is shown next to each answer choice. Of the remainders shown, only 2 is a prime. (1 is not prime.) A) 548 R4 B) 569 R1 C) 678 R6 D) 778 R2	25. D
26. My aunt can fold 4 paper cranes in 1 minute. My uncle can fold 3 paper cranes in 1 minute. Together they fold 7 paper cranes in 1 minute. It takes them $42 \div 7 = 6$ minutes to fold 42 paper cranes. A) 6 minutes B) 9 minutes C) 12 minutes D) 13 minutes	26. A
27. The second sum replaces 1 with 101, so the total is $2500 + 100 = 2600$. A) 2500 B) 2600 C) 2601 D) 2700	27. B
28. Work backwards. Alfonse's rat is $6 \times 4 = 24$ mm tall. His cat is $8 \times 24 = 192$ mm tall. His high chair is $10 \times 192 = 1920$ mm tall. A) 28 mm B) 480 mm C) 960 mm D) 1920 mm	28. D
29. If Ray ran for the first time last month on a Monday, then he ran on Wed., Fri., Sun., Tues., Thurs., Sat., Mon., Wed., and Fri. The tenth day was a Friday. A) Monday B) Tuesday C) Friday D) Sunday	29. C
30. Add 10 to 1, 3, 5, 7, . . . , 87, and 89. None of these sums is more than 99. There are 45 such sums. A) 45 B) 46 C) 90 D) 91	30. A

The end of the contest 4

Visit our Web site at http://www.mathleague.com

Math League Press, P.O. Box 17, Tenafly, New Jersey 07670-0017

Information & Solutions

Spring, 2014

4

Contest Information

- **Solutions** Turn the page for detailed contest solutions (written in the question boxes) and letter answers (written in the *Answer Column* to the right of each question).

- **Scores** Please remember that *this is a contest, and not a test*—there is no "passing" or "failing" score. Few students score as high as 24 points (80% correct); students with half that, 12 points, *deserve commendation!*

- **Answers and Rating Scales** Turn to page 140 for the letter answers to each question and the rating scale for this contest.

	Answers
1. Since 0 is a factor, $2 \times 0 \times 1 \times 4 = 0$. A) 0 B) 7 C) 8 D) 2014	1. A
2. Cubby just graduated, but his claws make it hard for him to hold his diploma! He has five claws on each of his four paws, for a total of $5 \times 4 = 20$ claws. A) 9 B) 16 C) 18 D) 20	2. D
3. If 2 years ago I was 3 years old, I am now $3 + 2 = 5$. A) 4 B) 5 C) 6 D) 23	3. B
4. I had \$1 and then I spent 55¢. Since 100¢ − 55¢ = 45¢, I have 45¢ left. A) 45¢ B) 55¢ C) 65¢ D) 75¢	4. A
5. $8 + (60 \div 4) = 8 + (15) = 23$. A) 15 B) 17 C) 22 D) 23	5. D
6. A summer vacation exactly 91 days long lasts for $91 \div 7 = 13$ weeks. A) 9 B) 11 C) 12 D) 13	6. D
7. $(1 + 7) + (2 + 6) + (3 + 5) = 3 \times 8 = 24 = 4 + 20$. A) 4 B) 20 C) 24 D) 28	7. B
8. 50 more than 400 is 450, and 3 less than 450 is 447. A) 87 B) 91 C) 447 D) 474	8. C
9. The prime numbers less than 10 are 2, 3, 5, and 7. A) 2 B) 3 C) 4 D) 5	9. C
10. Five nickels + ten dimes = 25¢ + 100¢ = 125¢ = 5 × 25¢ = 5 quarters. A) 5 B) 10 C) 15 D) 25	10. A
11. Caleb the dog dreams he has 12 dozen bones. Since 12 dozen = $12 \times 12 = 144$, there are $144 \div 2 = 72$ pairs. Caleb will have to dig 72 holes. A) 24 B) 72 C) 144 D) 288	11. B
12. Since 42 424 is between 42 400 and 42 500 but closer to 42 400, it rounds down to 42 400. A) 42 000 B) 42 100 C) 42 400 D) 42 500	12. C

Go on to the next page ⟫⟫➔ **4**

	Answers
13. From 9:45 PM to 10:45 PM is 60 mins. From 10:45 PM to 11 PM is 15 mins. From 11 PM to 11:10 PM is 10 mins. That's (60 + 15 + 10) mins. A) 65 B) 75 C) 85 D) 95	13. C
14. $124 \div 8 = 15R4$. A) 2 B) 4 C) 5 D) 6	14. B
15. $2014 \times 400 = 805\,600$; the hundreds digit is 6. A) 0 B) 5 C) 6 D) 8	15. C
16. I am going to travel 200 km at 60 km per hour. Since 60 km per hour is the same as 1 km per minute, my trip will take 200 minutes. A) 60 B) 200 C) 260 D) 320	16. B
17. Greta was 110 cm tall 2 years ago, when she was 10 cm taller than her brother. Her brother was 100 cm then. Now Greta is 10 cm shorter than her brother. If her brother grew 40 cm, he is now 140 cm. Greta is now 130 cm tall. A) 120 B) 130 C) 140 D) 150	17. B
18. $1000 \div 100 = 10 = 10 \times 1$. A) 1 B) 10 C) 100 D) 1000	18. A
19. From January 1st to January 31st, there are 16 odd-numbered dates. From February 1st to February 21st, there are 11 odd-numbered dates. That's 27 × \$2 = \$54. A) \$48 B) \$50 C) \$52 D) \$54	19. D
20. The whole numbers 8 and 12 are both divisible by 4. A) 1 B) 2 C) 3 D) 4	20. B
21. $9\times9 + 9\times8 + 9\times7 + 9\times6 = 9\times(9+8+7+6)$. A) 20 B) 24 C) 30 D) 36	21. C
22. Since $30 \div 3 = 10$, Chester put 10 groups of 3 pencils in his holder. Therefore, he chewed up 10 groups of 2 pencils. Chester chewed up $10 \times 2 = 20$ pencils. A) 12 B) 15 C) 20 D) 45	22. C

Go on to the next page ➠ **4**

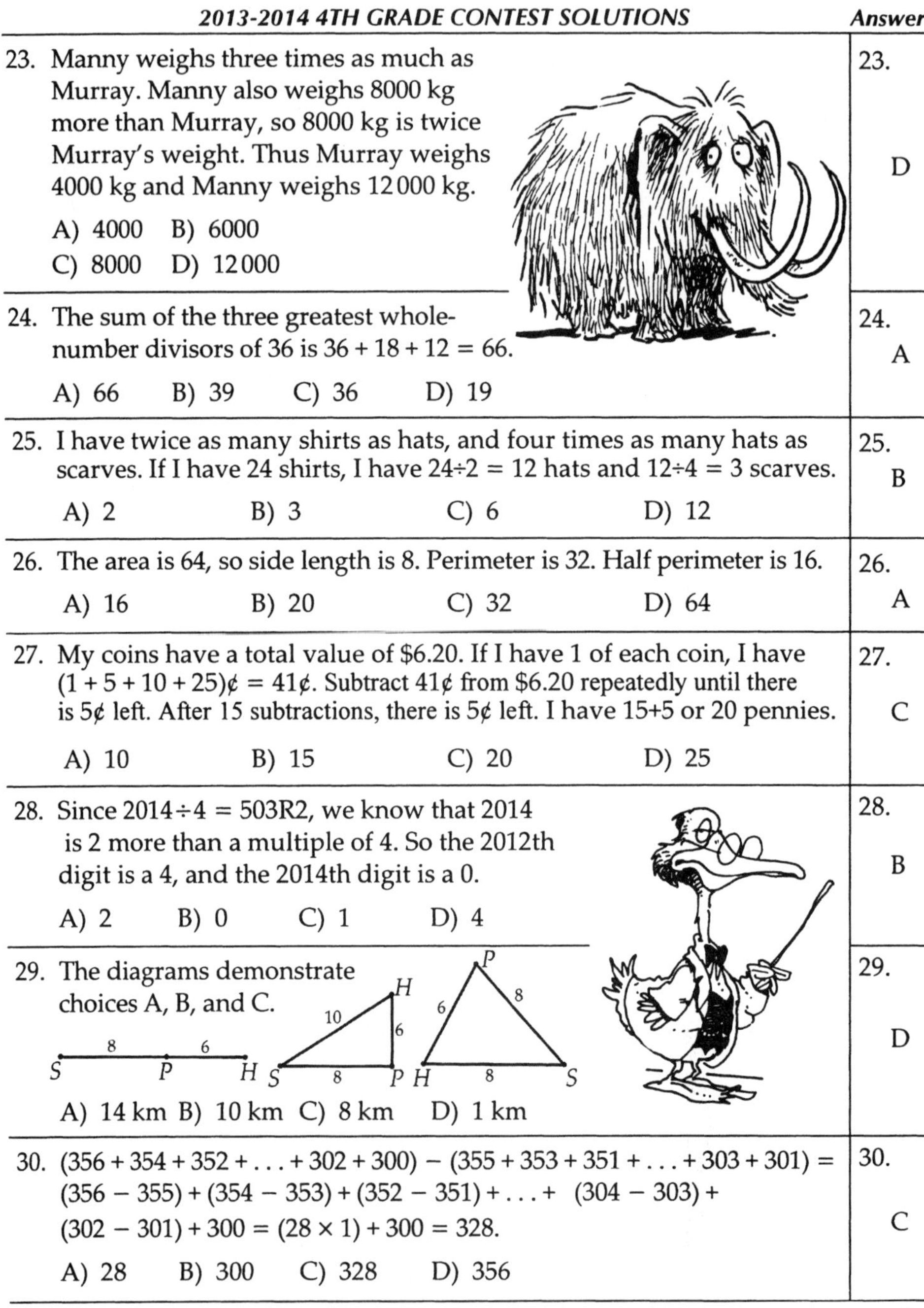

Solution	Answers
23. Manny weighs three times as much as Murray. Manny also weighs 8000 kg more than Murray, so 8000 kg is twice Murray's weight. Thus Murray weighs 4000 kg and Manny weighs 12 000 kg. A) 4000 B) 6000 C) 8000 D) 12 000	23. D
24. The sum of the three greatest whole-number divisors of 36 is $36 + 18 + 12 = 66$. A) 66 B) 39 C) 36 D) 19	24. A
25. I have twice as many shirts as hats, and four times as many hats as scarves. If I have 24 shirts, I have $24 \div 2 = 12$ hats and $12 \div 4 = 3$ scarves. A) 2 B) 3 C) 6 D) 12	25. B
26. The area is 64, so side length is 8. Perimeter is 32. Half perimeter is 16. A) 16 B) 20 C) 32 D) 64	26. A
27. My coins have a total value of \$6.20. If I have 1 of each coin, I have $(1 + 5 + 10 + 25)$¢ = 41¢. Subtract 41¢ from \$6.20 repeatedly until there is 5¢ left. After 15 subtractions, there is 5¢ left. I have 15+5 or 20 pennies. A) 10 B) 15 C) 20 D) 25	27. C
28. Since $2014 \div 4 = 503R2$, we know that 2014 is 2 more than a multiple of 4. So the 2012th digit is a 4, and the 2014th digit is a 0. A) 2 B) 0 C) 1 D) 4	28. B
29. The diagrams demonstrate choices A, B, and C. A) 14 km B) 10 km C) 8 km D) 1 km	29. D
30. $(356 + 354 + 352 + \ldots + 302 + 300) - (355 + 353 + 351 + \ldots + 303 + 301) =$ $(356 - 355) + (354 - 353) + (352 - 351) + \ldots + (304 - 303) +$ $(302 - 301) + 300 = (28 \times 1) + 300 = 328$. A) 28 B) 300 C) 328 D) 356	30. C

The end of the contest 4

Visit our Web site at http://www.mathleague.com

Math League Press, P.O. Box 17, Tenafly, New Jersey 07670-0017

Information & Solutions

Spring, 2015

4

Contest Information

- **Solutions** Turn the page for detailed contest solutions (written in the question boxes) and letter answers (written in the *Answer Column* to the right of each question).

- **Scores** Please remember that *this is a contest, and not a test*—there is no "passing" or "failing" score. Few students score as high as 24 points (80% correct); students with half that, 12 points, *deserve commendation!*

- **Answers and Rating Scales** Turn to page 141 for the letter answers to each question and the rating scale for this contest.

2014-2015 4TH GRADE CONTEST SOLUTIONS	Answers
1. $2 \times 14 = 28 = 4 \times 7$. A) 7 B) 10 C) 20 D) 28	1. A
2. Each nilbog has 2 hands with 6 fingers and 2 feet with 6 toes. If there are 3 nilbogs, they have $3 \times 6 = 18$ fingers and $3 \times 6 = 18$ toes. They have a total of $18 + 18 = 36$ fingers and toes.  A) 9 B) 12 C) 18 D) 36	2. D
3. $(10 + 20) \times (30 + 40) = 30 \times 70 = 2100$. A) 100 B) 650 C) 940 D) 2100	3. D
4. The number that is 34 more than 56 is $56 + 34 = 90$. A) 90 B) 80 C) 32 D) 22	4. A
5. $7 \times (8 \times 9 \times 10 \times 11) = (8 \times 9 \times 10 \times 11) \times 7$. A) 5 B) 7 C) 12 D) 18	5. B
6. 9876 is between 9870 and 9880; it is closer to 9880 than 9870. A) 9800 B) 9870 C) 9880 D) 9900	6. C
7. The perimeter is $4 + 9 + 4 + 9 = 26$. 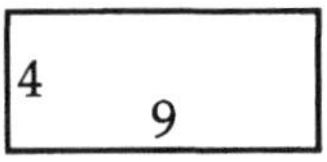 A) 13 B) 26 C) 36 D) 49 (rectangle: 4, 9)	7. B
8. 5 quarters = \$1.25. 5 nickels = 25¢. We need \$1 more or 10 dimes. A) 5 B) 10 C) 15 D) 20	8. B
9. The sum of 123 and 345 is 468. Since 8 is even, 468 is even. A) even B) odd C) prime D) over 500	9. A
10. The factors of 20 are 1, 2, 4, 5, 10, and 20. A) 10 B) 15 C) 20 D) 25	10. C
11. Quincy Magoo Jr. led a hike. The map said to walk 14×32 steps = 448 steps. Quincy walked 12×34 steps = 408 steps. He walked $448 - 408 = 40$ fewer steps than the map said to.  A) 4 B) 12 C) 16 D) 40	11. D
12. 7 days from Tues. is Tues.; 3 more is Fri. A) Sunday B) Monday C) Friday D) Saturday	12. C

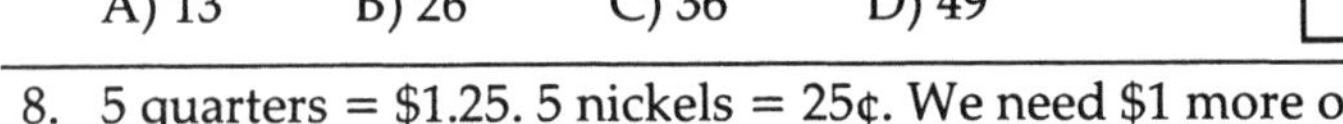

13. Father Time is jumping up and down to celebrate a new year. On jump 1 he jumps 10 cm, on jump 2 he jumps 20 cm, on jump 3 he jumps 40 cm, jump 4 is 80 cm , and jump 5 is 160 cm = 1.6 m. His first jump of over 1 m is jump 5.

A) 4 B) 5 C) 10 D) 11

13. B

14. (88 + 88 + 88) ÷ 8 = 11 + 11 + 11 = 33 = 99 ÷ 3.

A) 3 B) 9 C) 11 D) 33

14. A

15. Use ones digits only: the ones digit of 4×5+5×6+6×7+7×8 is 2+6 = 8.

A) 0 B) 2 C) 6 D) 8

15. D

16. I will be 5 years older than I am today in 5 years.

A) 5 B) 8 C) 11 D) 24

16. A

17. Try starting with 2. Triple it to get 6 and then triple 6 to get 18. Now, subtract 3 to get 15. The result is always divisible by 3.

A) 9 B) 6 C) 3 D) 2

17. C

18. I spent 6 × \$1.05 = \$6.30. You spent 8 × 90¢ = \$7.20. I spent \$7.20 − \$6.30 = \$0.90 less than you did.

A) \$0.90 B) \$1.00 C) \$1.10 D) \$1.20

18. A

19. List C puts the 3 shapes in order from most sides to least sides since the figures have 5, 4, and 3 sides respectively.

A) triangle, square, pentagon B) square, triangle, rectangle
C) pentagon, rectangle, triangle D) pentagon, triangle, square

19. C

20. Tom painted 7 walls earlier today. After he completes the one he is painting now, he will have painted 8 walls. At the end of the day he will have painted 8 + 11 = 19 walls.

A) 18 B) 19 C) 20 D) 21

20. B

21. (4 × 444) × 2 = 4 × (444 × 2) = 4 × 888.

A) 444 B) 666 C) 888 D) 1616

21. C

22. There are 2 × 5 = 10 fingers on a pair of gloves. A dozen pairs have 12 × 10 = 120.

A) 24 B) 30 C) 60 D) 120

22. D

Go on to the next page ⟫➡ **4**

23. At the end of the first half, the Roaring Hamsters were 1 point behind the Screaming Gerbils. During the 2nd half, the Gerbils didn't score, and the Hamsters finished ahead by 7, so the Hamsters scored 8 points during the second half. In all, the Hamsters scored 16 points.

    A) 16 B) 14 C) 12 D) 8

    23. A

24. $1+2+3+4+5+6+7+8+9+10$ is 55. Since Jon's sum was 48, he left out $55-48=7$.

    A) 6 B) 7 C) 8 D) 9

    24. B

25. 10 000 000 has 2 more zeros than 100 000.

    A) 1 B) 2 C) 3 D) 4

    25. B

26. If a number is divided by 100, the remainder is the number's last 2 digits.

    A) 21 B) 42 C) 65 D) 98

    26. A

27. My old printer printed 13 pages in 3 seconds or 26 pages in 6 seconds. My new printer prints 21 pages in 2 seconds or 63 pages in 6 seconds. The difference is $63-26=37$.

    A) 8 B) 16 C) 27 D) 37

    27. D

28. The ones digit of the product is the same as the ones digit of 0^4, 1^4, 2^4, 3^4, 4^4, 5^4, 6^4, 7^4, 8^4, or 9^4. The ones digit can be 0, 1, 5, or 6.

    A) 1 B) 4 C) 5 D) 6

    28. B

29. Any number divisible by 9 and 10 is divisible by 90. The divisors of 90 are 1, 2, 3, 5, 6, 9, 10, 15, 18, 30, 45, and 90. Since 90 has 12 divisors, the number is divisible by 12 whole numbers.

    A) 6 B) 8 C) 10 D) 12

    29. D

30. The ones digits in their ages may be 5 & 0, 6 & 1, 7 & 2, 8 & 3, or 9 & 4. The sum of these two digits will have a ones digit of 5, 7, 9, 1, or 3.

    A) 42 B) 56 C) $63=34+29$ D) 78

    30. C

The end of the contest  4

Visit our Web site at http://www.mathleague.com

Math League Press, P.O. Box 17, Tenafly, New Jersey 07670-0017

Information & Solutions

Spring, 2016

4

Contest Information

- **Solutions** Turn the page for detailed contest solutions (written in the question boxes) and letter answers (written in the *Answer Column* to the right of each question).

- **Scores** Please remember that *this is a contest, and not a test*—there is no "passing" or "failing" score. Few students score as high as 24 points (80% correct); students with half that, 12 points, *deserve commendation!*

- **Answers and Rating Scales** Turn to page 142 for the letter answers to each question and the rating scale for this contest.

Solution	Answers
1. $4 \times 9 = 36$. A) $16 \times 2 = 32$ B) $12 \times 3 = 36$ C) $7 \times 5 = 35$ D) $38 \times 1 = 38$	1. B
2. 3 forks cost $\$6 \times 3 = \$18$. 4 spoons cost $\$7 \times 4 = \$28$. 5 knives cost $\$8 \times 5 = \$40$. They cost a total of $\$18 + \$28 + \$40 = \$86$. A) \$40 B) \$46 C) \$80 D) \$86	2. D
3. In division, the remainder must be less than the divisor. When a number is divided by 5, the remainder cannot be 5. A) 0 B) 1 C) 3 D) 5	3. D
4. The perimeter of a rectangle is $2\times(\text{length} + \text{width}) = 2\times(20 + 16) = 72$. A) 320 B) 160 C) 72 D) 36	4. C
 5. $10000 \div 200 = 50$; $50 \times \underline{200} = 10000$. A) 100 B) 200 C) 1000 D) 2000	5. B
6. The greatest common factors for the choices are 1, 3, 1, 1, respectively. A) 4 and 9 B) 6 and 27 C) 27 and 50 D) 33 and 100	6. B
7. 5 m + 5 cm + 5 mm = 5000 mm + 50 mm + 5 mm = 5055 mm. A) 5055 B) 5505 C) 5550 D) 55550	7. A
8. 33 hours and 36 minutes = $33 \times 60 + 36$ minutes = 2016 minutes. A) 1996 B) 2006 C) 2016 D) 2026	8. C
9. Since $3 \times 3 = 9$, the product of two multiples of 3 is divisible by 9. A) 6 B) 9 C) 12 D) 15	9. B
10. The whole-number factors of 49 are 1, 7, and 49. A) 47 B) 48 C) 49 D) 50	10. C
11. After selling 1 book and buying 2 books, Joe had 31 books. After selling 3 books and buying 4 books, Joe had 32 books. After selling 5 books and buying 6 books, Joe had 33 books. A) 9 B) 27 C) 33 D) 51	11. C

Go on to the next page ⟫➡ **4**

Solutions	Answers
12. On Day 2 there are 128 apples. On Day 3 there are 64 apples. On Day 4 there are 32 apples. On Day 5 there are 16 apples. On Day 6 there are 8 apples. On Day 7 there are 4 apples. On Day 8 there are 2 apples. On Day 9 there is only 1 apple. A) 10 B) 9 C) 8 D) 7	12. B
13. The perimeter of a square is 4 multiplied by the side-length. The area of a square is the side-length multiplied by the side-length. A) 1 B) 2 C) 4 D) 16	13. C
14. The ones digit of $2015^{2015} + 2016^{2016}$ is the same as that of $5+6$. A) 1 B) 3 C) 5 D) 7	14. A
15. $24 \times 26 \times 28 \times 30 \times 32 = (24 \times 26 \times 28 \times 30) \times (2 \times 2 \times 2 \times 2) \times \underline{2}$. A) 2 B) 4 C) 16 D) 64	15. A
16. 12000 hours after 3 P.M. is 3 P.M. 11999 hours after 3 P.M is 2 P.M. A) 2 A.M. B) 2 P.M. C) 4 A.M. D) 4 P.M.	16. B
17. The sums of the digits of both numbers are divisible by 9. A) 2 B) 7 C) 9 D) 11	17. C
18. The product of a whole number and an even number must be even. A) prime B) composite C) odd D) even	18. D
19. My age now is $15 + 15 = 30$. In 15 years I will be $30 + 15 = 45$. A) 15 B) 30 C) 45 D) 60	19. C
20. I spent a total of 45 min. $\times 4 = 180$ min. on English assignments. I spent a total of 40 min. $\times 6 = 240$ min. on math assignments. I spent a total of $180 + 240 = 420$ min., which is 7 hours, on homework assignments. A) 6 B) 7 C) 420 D) 430	20. B
21. $2016 \times (20 + 40 + 60) = 2016 \times 120$. A) 120 B) 240 C) 2016 D) 48000	21. A
22. 365 days have $365 \times 24 \times 60 \times 60 \div 1000 = 31536$ thousand seconds. A) 31536 B) 525600 C) 1892160 D) 3536000	22. A

Go on to the next page ⟹ 4

2015-2016 4TH GRADE CONTEST SOLUTIONS	Answers
23. By noon, Tom had caught 47 more fish than Jerry had by 1 P.M. The total number of fish Jerry caught was 19 more than the total number that Tom caught. Jerry caught $47 + 19 = 66$ more fish after noon than he caught before. If catching 66 more fish tripled Jerry's total, then the 66 more fish **added** twice as many fish to his total. Jerry had 33 fish at noon and a total of 99 fish by the end of the day. A) 33 B) 80 C) 99 D) 102	23. C
24. Each whole number in the second sequence is 5 more than the corresponding number in the first sequence. We have $280 + 5 \times 5 = 305$. A) 285 B) 305 C) 405 D) 425	24. B
25. Add the 6 zeros ending 1 million to the 9 zeros ending 1000 million. A) 15 B) 16 C) 17 D) 54	25. A
26. $1 \times 2 \times 3 \times 4 \times 5 \times 6 \times 7 \times 8 \times 9 \times 10$ is divisible by 3 and 25. A) 71 B) 73 C) 75 D) 77	26. C
27. 7 out of 11 balls that Elena bought are soccer balls. If there are $7 \times 5 =$ 35 soccer balls, then there is a total of $11 \times 5 = 55$ balls. A) 20 B) 35 C) 45 D) 55	27. D
28. The prime is < 33, so it could be 2, 3, 5, 7, 11, 13, 17, 19, 23, 29, or 31. A) 9 B) 10 C) 11 D) 12	28. C
29. Because the least common multiple of 3, 4, and 5 is 60, Barb does all 3 chores on the same day every 60 days. Since the remainder of $60 \div 7$ is 4, 60 days after Sunday is a Thursday. A) Wednesday B) Thursday C) Friday D) Saturday	29. B
30. $2016 = 2 \times 2 \times 2 \times 2 \times 2 \times 3 \times 3 \times 7$; there are 3 different primes. A) 3 B) 4 C) 7 D) 8	30. A

The end of the contest 4

Visit our Web site at http://www.mathleague.com

5th Grade Solutions

2011-2012 through 2015-2016

FIFTH GRADE MATHEMATICS CONTEST

Math League Press, P.O. Box 17, Tenafly, New Jersey 07670-0017

Information & Solutions

Spring, 2012

5

Contest Information

- **Solutions** Turn the page for detailed contest solutions (written in the question boxes) and letter answers (written in the *Answer Column* to the right of each question).

- **Scores** Please remember that *this is a contest, and not a test*—there is no "passing" or "failing" score. Few students score as high as 24 points (80% correct); students with half that, 12 points, *deserve commendation!*

- **Answers and Rating Scales** Turn to page 143 for the letter answers to each question and the rating scale for this contest.

	Answers
1. The forklift lifts 2012 kg at a time. In 4 lifts, it will lft 4 × 2012 kg = 8048 kg. A) 503 B) 2016 C) 4024 D) 8048	1. D
2. (22 + 44 + 66) ÷ (2 + 4 + 6) = 132 ÷ 12 = 11. A) 10 B) 11 C) 33 D) 44	2. B
3. The sum, difference, and product of even numbers are even, but 952 ÷ 136 = 7 is odd. A) 952 + 136 B) 952 − 136 C) 952 ÷ 136 D) 952 × 136	3. C
4. Since 100 ÷ 3 = 33 with remainder 1, it takes my frog 33 jumps to jump 99 m. It needs 1 more jump, for 34 jumps in all, to go at least 100 m. A) 33 B) 34 C) 70 D) 97	4. B
5. 100 × 100 + 10 × 10 + 1 × 1 = 10000 + 100 + 1 = 10101. A) 111 B) 1101 C) 1011 D) 10101	5. D
6. Every 4 quarters is $1; the number of dollars I have is 2500÷4 = $625. A) $100 B) $500 C) $625 D) $1000	6. C
7. It's 92 days (30+31+31) from Nov. 6 to Feb. 6; 8 days later is 100 days. A) February 14 B) February 15 C) February 16 D) February 17	7. A
8. Add ones digits, then divide by 10: 2+4+6+9+0 = 21; 21÷10 = 2 R 1. A) 0 B) 1 C) 4 D) 9	8. B
9. Connie counts from 1 to 20. The sum of the prime numbers she counts is 2 + 3 + 5 + 7 + 11 + 13 + 17 + 19 = 77. (Note: 1 is not prime.) A) 29 B) 30 C) 77 D) 78	9. C
10. Since (888888 + 888) ÷ 8 = 111111 + 111 = 111222, Flash has taken 111222 photos. A) 111111 B) 111222 C) 888999 D) 111111111	10. B
11. From 9:30 A.M. until 1:15 P.M. is 3 hours and 45 minutes. Add 3 hours and 45 minutes to 1:15 P.M. to get 5:00 P.M. A) 4:30 P.M. B) 5:00 P.M. C) 5:30 P.M. D) 9:30 P.M.	11. B

Go on to the next page ⟫⟫➔ 5

Solution	Answers
12. Inspector Ivan is looking for the largest number that is a factor of both 36 and 90. Since $36 = 2 \times 18$ and $90 = 5 \times 18$, it's 18. A) 2 B) 3 C) 9 D) 18	12. D
13. I drove 60 km in 20 min, an average of 3×60 km in 3×20 min = 180 km per hr. A) 20 B) 30 C) 120 D) 180	13. D
14. Since 10 is a factor of this product, the ones digit is 0. A) 0 B) 2 C) 5 D) 7	14. A
15. Each side of a square with perimeter 36 has length $36 \div 4 = 9$. A rectangle with two sides twice this length and 2 sides triple this length has dimensions 18 and 27. Its perimeter is $2 \times (18 + 27) = 90$. A) 60 B) 90 C) 216 D) 486	15. B
16. If 60 beans cost \$3.00, then 1 bean costs 5¢ and 100 beans cost \$5. A) \$3.50 B) \$4.00 C) \$5.00 D) \$5.50	16. C
17. My age, 27, is 3 years less than 5 times my sister's age, so 30 is 5 times her age. Thus, my sister is 6. The sum of our ages is $27 + 6 = 33$. A) 33 B) 36 C) 39 D) 42	17. A
18. I ate lunch every day of my 12-week summer break. I ate lunch outdoors only on every Saturday, Sunday, and Tuesday. I ate lunch indoors 4 days each week. In 12 weeks, that's $12 \times 4 = 48$ days. A) 24 B) 36 C) 48 D) 60	18. C
19. The average of 4000 fours is 4; this is 4 times the average of 2000 ones. A) ones B) twos C) fours D) eights	19. A
20. Quadrilaterals, trapezoids, and parallelograms each have 4 sides each, so the total number of sides is $4 + 4 + 4 = 12$. A) 12 B) 13 C) 14 D) 15	20. A
21. If Wally averaged 30 guests per day, that's 120 in 4 days. Since he had 58 guests the first 2 days, he had $120 - 58 = 62$ guests the last 2 days. He had equal numbers of guests the last 2 days, so he had 31 each day. A) 29 B) 30 C) 31 D) 32	21. C

Go on to the next page ⟹ **5**

	Answers
22. If they were all the same age, they'd each be $165 \div 3 = 55$. Instead, Gene is 55 years old, Emily is 50 years old, and Gene's car is 60 years old. A) 50 B) 55 C) 60 D) 65	22. C
23. When 14 (or any other even multiple of 7) is divided by 7, the remainder is 0. A) 0 B) 1 C) 5 D) 6	23. A
24. Divide through by 2 then multiply by 7 to get 4:14 = 2:7 = 14:49. A) 4 B) 24 C) 49 D) 114	24. C
25. The almanac, the dictionary, the biography, and the cookbook together weigh 4310 g + 2325 g = 6635 g. Since the almanac and the biography together weigh 2795 g, the cookbook and the dictionary together weigh 6635 g − 2795 g = 3840 g. A) 2620 g B) 3270 g C) 3840 g D) 4100 g	25. C
26. $(100-98)+(96-94)+\ldots+(8-6)+(4-2)+0=2\times 25+0=50$. A) 26 B) 50 C) 52 D) 100	26. B
27. Since $40\times 6=240$, 240 campers have signed up for the 6 sports, and each camper is counted 3 times. There are $240\div 3=80$ campers in all. A) 80 B) 120 C) 160 D) 180	27. A
28. There are 24 people behind me and 24 in front of my friend. That's a total of 48 people. But we counted the 10 people between me and my friend twice. Thus, there are $48-10=38$ people on line. A) 37 B) 38 C) 56 D) 58	28. B
29. Carl's dad can feed 12 babies for 21 days. That's like feeding $12\div 4=3$ babies for $21\times 4=84$ days. Therefore, he could feed $3\times 3=9$ babies for $84\div 3=28$ days. A) 16 B) 18 C) 24 D) 28	29. D
30. The sum must be a multiple of 3. There are many examples, e.g., 2 & 1, 4 & 2, 10 & 5, etc. The quotient is always 2. A) 2 B) 3 C) 4 D) 5	30. A

The end of the contest 5

Visit our Web site at http://www.mathleague.com

FIFTH GRADE MATHEMATICS CONTEST

Math League Press, P.O. Box 17, Tenafly, New Jersey 07670-0017

Information & Solutions

Spring, 2013

Contest Information

5

- **Solutions** Turn the page for detailed contest solutions (written in the question boxes) and letter answers (written in the *Answer Column* to the right of each question).

- **Scores** Please remember that *this is a contest, and not a test*—there is no "passing" or "failing" score. Few students score as high as 24 points (80% correct); students with half that, 12 points, *deserve commendation!*

- **Answers and Rating Scales** Turn to page 144 for the letter answers to each question and the rating scale for this contest.

1. Since 14 days before Saturday is Saturday, 4 more days before that would be Tuesday.
A) Tuesday B) Wednesday
C) Thursday D) Friday

Answer 1. A

2. $(1+2+3)\times 10 = 60 = 30+20+10$.
A) 10 B) 11 C) 33 D) 44

Answer 2. A

3. I listened to 6 songs before the one I'm listening to now, and I will listen to 6 more after this one. That's $6+1+6 = 13$ songs.
A) 11 B) 12 C) 13 D) 14

Answer 3. C

4. 100 hundreds ÷ 10 tens = $10000 \div 100 = 100$.
A) 10 B) 100 C) 1000 D) 10000

Answer 4. B

5. $9+99+999 = 9\times(1+11+111) = 9\times 123$.
A) 111 B) 112 C) 122 D) 123

Answer 5. D

6. I created 30 characters, 3 for each video game I own. That means I own $30\div 3 = 10$ video games.
A) 10 B) 33 C) 40 D) 90

Answer 6. A

7. If I add the number of sides that a hexagon has (6) to the number of sides that a pentagon has (5), then the sum is $6+5 = 11$, which is odd.
A) rhombus B) square C) pentagon D) quadrilateral

Answer 7. C

8. $40+30\times 20+10\times 0 = 40+600+0 = 640$.
A) 0 B) 150 C) 640 D) 1400

Answer 8. C

9. Subtract 6 from 30 to get 24, which is twice my age. Therefore, I am 12 years old. My brother is 6 years older than I am, so he is 18.
A) 12 B) 15 C) 18 D) 21

Answer 9. C

10. Since \$50 − \$16 = \$34, Don paid \$34 ÷ 5 = \$6.80 per tropical punch.
A) \$5.20 B) \$6.80
C) \$8.20 D) \$8.80

Answer 10. B

11. The average of 12 and 24 is $(12+24)\div 2 = 18$.
A) 13 B) 18 C) 24 D) 36

Answer 11. B

Go on to the next page ⟫⟫➜ 5

	Answers
12. One hr. and 46 min. = (60 + 46) min. = 106 minutes. Playing at twice that speed, it would take Manuel 106 ÷ 2 = 53 minutes to play the concerto. A) 53 B) 73 C) 83 D) 212	12. A
13. Add 60 to 180 and divide by 3: 240 ÷ 3 = 80. A) 40 B) 60 C) 70 D) 80	13. D
14. There are a total of 2013 students enrolled at 8 high schools. There are 234 students at each of 4 of the schools, for a total of 936 students. That leaves 2013 − 936 = 1077 students. A) 1077 B) 1123 C) 1234 D) 1443	14. A
15. Three different books, *A, B, C,* are arranged on my bookshelf. They may be arranged as *ABC, ACB, BAC, BCA, CAB,* or *CBA.* A) 3 B) 4 C) 5 D) 6	15. D
16. A square piece of paper has a perimeter of 36 cm. Twice the perimeter is 72 cm. Each side is 72÷4 = 18 cm, and the area is 324 cm^2. A) 72 cm^2 B) 108 cm^2 C) 144 cm^2 D) 324 cm^2	16. D
17. The value of 1 quarter, 1 dime, and 1 nickel is 40¢. My coins must have a total value divisible by 40, but $3.80 is not divisible by 40. A) $2.40 B) $3.80 C) $4.40 D) $5.20	17. B
18. The whole number factors of 12 are 1, 2, 3, 4, 6, and 12. A) 6 B) 9 C) 12 D) 16	18. C
19. The least common multiple of 10 and 24 is 120; the greatest common factor of 10 and 24 is 2. Their sum is 120 + 2 = 122. A) 121 B) 122 C) 241 D) 242	19. B
20. For every 8 vehicles in the lot, 5 are cars and 3 are trucks. If the lot has 120 vehicles, that's 15 groups of 8. Each group has 5 cars: 15 × 5 = 75. A) 24 B) 45 C) 75 D) 80	20. C
21. A rate of 600 m/min. = 60 000 cm/min. = 60 000 cm/ 60 sec. = 1000 cm/sec. A) 100 B) 600 C) 1000 D) 60000	21. C

Go on to the next page ⟫⟫➡ 5

Solution	Answers
22. Maria had 28 dreams last month, 24 of which involved animals. Since 16 + 15 = 31 involved moneys or squirrels, then at least 31 − 24 = 7 dreams involved both monkeys and squirrels. A) 3 B) 7 C) 9 D) 11	22. B
23. A trapezoid may have consecutive sides of lengths 3, 3, 8, and 4. A) triangle B) square C) parallelogram D) trapezoid	23. D
24. My pennies can be divided into 500÷6 = 83 groups of 6 pennies, with 2 left over. At the end of the 82nd day, I will have 6 + 2 = 8 pennies left. A) 6 B) 8 C) 10 D) 12	24. B
25. A fair sells a "combo" ticket for $30 entry and a "per ride" ticket for $12.50 to enter plus $5 per ride. A "per ride" ticket costs $12.50 + $15 = $27.50 for 3 rides and $12.50 + $20 = $32.50 for 4 rides. A) 3 B) 4 C) 6 D) 7	25. B
26. Since 5 × 4 = 20, the ones digit of the given product must be 0. A) 0 B) 4 C) 6 D) 9	26. A
27. A team scores an average of 31 points per game in its 1st 4 games for a total of 31 × 4 = 124 points, and an average of 30 points per game in its 1st 5 games for a total of 30 × 5 = 150 points. The difference is 150 − 124 = 26. A) 26 B) 27 C) 28 D) 29	27. A
28. The possibilities are 1) XOXOXOO, 2) XOXOOXO, 3) XOXOOOX, 4) XOOXOXO, 5) XOOXOOX, 6) XOOOXOX, 7) OXOXOXO, 8) OXOXOOX, 9) OXOOXOX, and 10) OOXOXOX. A) 4 B) 6 C) 8 D) 10	28. D
29. Mo and Jo with Bo and Ko have a total of 273 coins. If we subtract the 127 coins Mo and Bo have, Jo and Ko have 146. A) 106 B) 128 C) 135 D) 146	29. D
30. Place 100 2 x 2 squares in a line. The perimeter is 2 x (2 + 200) = 404. A) 88 B) 100 C) 400 D) 404	30. D

The end of the contest 5

Visit our Web site at http://www.mathleague.com

FIFTH GRADE MATHEMATICS CONTEST

Math League Press, P.O. Box 17, Tenafly, New Jersey 07670-0017

Information & Solutions

Spring, 2014

5

Contest Information

- **Solutions** Turn the page for detailed contest solutions (written in the question boxes) and letter answers (written in the *Answer Column* to the right of each question).

- **Scores** Please remember that *this is a contest, and not a test*—there is no "passing" or "failing" score. Few students score as high as 24 points (80% correct); students with half that, 12 points, *deserve commendation!*

- **Answers and Rating Scales** Turn to page 145 for the letter answers to each question and the rating scale for this contest.

Solution	Answers
1. Larry's bird wished him a happy graduation by calling in his ear, singing 4 notes each time it called. The bird called 8 times, waited, and then called 4 more times. It sang a total of $4 \times (8+4) = 48$ notes. A) 3 B) 12 C) 16 D) 48	1. D
2. $(2014-1014)+(3014-2014) = 1000+1000 = 2000$. A) 0 B) 1000 C) 2000 D) 2014	2. C
3. The sum of the measures in degrees of two of the angles in a rectangle is $90+90 = 180$. A) 90 B) 180 C) 200 D) 360	3. B
4. $10+(9\times 8)-(7\times 6) = 10+72-42 = 40$. A) 40 B) 110 C) 114 D) 870	4. A
5. The smallest difference is $2014-1983 = 31$. A) 1914 B) 1983 C) 2056 D) 3014	5. B
6. The prime factorization of 72 is $2\times2\times2\times3\times3$. The largest prime is 3. A) 3 B) 7 C) 36 D) 72	6. A
7. Sam drives three times as fast as Dave. If Dave drives 200 km in 6 hours, Sam drives 200 km in $6\div 3 = 2$ hours. A) 2 B) 4 C) 12 D) 18	7. A
8. $6\times 4 = 24 = 96\div 4$. A) 6 B) 12 C) 24 D) 96	8. D
9. From 8:18 AM to 12:18 PM is 4 hrs. = 240 min. From 12:18 PM to 2:18 PM is 2 hrs. = 120 min. From 2:18 PM to 2:36 PM is 18 minutes. That's 378 minutes in all. A) 258 B) 378 C) 418 D) 618	9. B
10. If 6 cans contain 96 teaspoons of sugar, 1 can contains $96\div 6 = 16$ teaspoons of sugar. Thus 15 cans contain $16\times 15 = 240$ teaspoons of sugar. A) 192 B) 208 C) 240 D) 288	10. C
11. $88\times 88 = (8\times 11)\times(8\times 11) = 8\times(11\times 11\times 8)$. A) 11×2 B) $11+11$ C) 11×11 D) $11\times 11\times 8$	11. D
12. The largest possible such sum is $98+99 = 197$. A) 21 B) 99 C) 197 D) 198	12. C

Go on to the next page ⟫ 5

	Answers
13. On 6 days of a certain week, Alexander's ragtime band played 2 + 16 + 3 + 15 + 4 + 14 = 54 songs. The band averaged 9 songs per day for a total of 7×9 = 63 songs for the entire week. The band played 63 − 54 = 9 songs on the 7th day of the week. A) 5 B) 7 C) 9 D) 11	13. C
14. The number 789 678 567 456 is added to the number 987 876 765 654. Since we carry a 1 when adding the left-most digits, the sum has 12 + 1 digits. A) 12 B) 13 C) 24 D) 25	14. B
15. Since \$20 − \$4.44 = \$15.56, four orders of French fries totaled \$15.56. To find the cost of one order, divide this total by 4: \$15.56 ÷ 4 = \$3.89. A) \$1.39 B) \$3.89 C) \$3.92 D) \$7.78	15. B
16. We must find which number among the choices is two more than a multiple of 5. Divide each choice by 5 (or recognize that any number that ends in "2" or "7" is 2 more than a multiple of 5). A) 4351 B) 5215 C) 5616 D) 6462	16. D
17. The ones digit is the same as the ones digit of 6 × 7 × 8 × 9 × 0. A) 0 B) 2 C) 4 D) 8	`17. A
18. Of every 11 people, there are 2 adults and 9 children. Since 99÷11 = 9, there are 9 groups of 11 people. Of these, 9×2 = 18 are adults. A) 9 B) 11 C) 18 D) 22	18. C
19. (# of sides a rhombus has) × (# of sides a square has) = 4 × 4 = 16 = (# of sides a rectangle has) × (# of sides a trapezoid has). A) triangle B) trapezoid C) hexagon D) octagon	19. B
20. Ann sent Wilson hearts with odd numbers with odd tens digits. The number on each heart he received must be two digits with both digits odd. There are 5 possible tens digits and 5 possible ones digits. That's a total of 5 × 5 = 25 hearts. A) 23 B) 25 C) 30 D) 45	20. B
21. 25×4 = 100; the only non-zero # in the product of 25 000 000 and 40 000 000 is the 1. A) 4 B) 3 C) 2 D) 1	21. D

Go on to the next page ⟫⟫➡ 5

22. Since Rich ate his favorite sandwich 8 days ago, today is the 9th day of the month. Since the shortest month has 28 days it is at least 28 − 9 = 19 days until the last day of the month. He must wait 1 more day.

A) 19 B) 20 C) 22 D) 23

22. B

23. Work backwards: 99 ÷ 3 = 33, 33 − 3 = 30, and 30 ÷ 3 = 10. I am now 10. Three years ago I was 7.

A) 7 B) 10 C) 13 D) 21

23. A

24. The factors of 49 are 1, 7, and 49. Since 49 has 3 factors, it has a prime number of factors.

A) 6 B) 12 C) 36 D) 49

24. D

25. The girls must be in the odd-numbered seats; otherwise, two girls would be seated next to each other. This leaves only the even-numbered seats for the boys, so a boy is seated in seat 4.

A) 1 B) 3 C) 4 D) 5

25. C

26. Dividing a certain two-digit number by 10 leaves a remainder of 9, so it is 19, 29, 39, 49, 59, 69, 79, 89, or 99. The only number listed with remainder 8 when divided by 9 is 89, so the number is 89 and 8 + 9 = 17.

A) 7 B) 9 C) 13 D) 17

26. D

27. The rectangle's longer sides are 3 times as long as the shorter sides. Half the perimeter is 12, so the rectangle has dimensions 3 by 9 and area 27.

A) 18 B) 24 C) 27 D) 48

27. C

28. The whole numbers less than 1000 that can be written as such a product are 0×1×2, 1×2×3, 2×3×4, 3×4×5, 4×5×6, 5×6×7, 6×7×8, 7×8×9, 8×9×10, and 9×10×11. In all, that's 10.

A) 10 B) 11 C) 15 D) 21

28. A

29. As shown, the greatest number of 4 by 6 rectangles that can be drawn on Bill's canvas is 5.

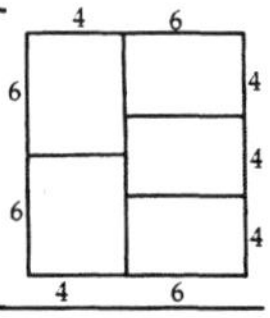

A) 2 B) 3 C) 4 D) 5

29. D

30. The only such numbers are 5432, 5431, 5430, 5421, 5420, 5410, 5321, 5320, 5310, and 5210. In all, there are 10 such numbers.

A) 3 B) 10 C) 69 D) 120

30. B

The end of the contest **5**

Visit our Web site at http://www.mathleague.com

Math League Press, P.O. Box 17, Tenafly, New Jersey 07670-0017

Information & Solutions

Spring, 2015

5

Contest Information

- **Solutions** Turn the page for detailed contest solutions (written in the question boxes) and letter answers (written in the *Answer Column* to the right of each question).

- **Scores** Please remember that *this is a contest, and not a test*—there is no "passing" or "failing" score. Few students score as high as 24 points (80% correct); students with half that, 12 points, *deserve commendation!*

- **Answers and Rating Scales** Turn to page 146 for the letter answers to each question and the rating scale for this contest.

	Answers
1. Mary Hartman has been shouting for 3 hours. She has been shouting for $3 \times 60 = 180$ minutes. A) 90 B) 150 C) 180 D) 300	1. C
2. $6 + (6 \div 6) + (6 \times 6) - 6 = 6 + 1 + 36 - 6 = 37$. A) 37 B) 42 C) 43 D) 48	2. A
3. Since $2015 = 2014 + 1$, the remainder is 1. A) 0 B) 1 C) 4 D) 2013	3. B
4. The perimeter of this hexagon $6 \times 6 = 36$. A) 12 B) 24 C) 30 D) 36	4. D
5. $12 \times 12 = 12 \times (3 \times 4) = (12 \times 3) \times 4 = 36 \times 4$. A) 4 B) 6 C) 9 D) 36	5. A
6. A hamburger costs \$5.50 and a soda costs \$0.75. The total cost of 3 hamburgers and 2 sodas is 3×\$5.50 + 2×\$0.75 = \$16.50 + \$1.50 = \$18. A) \$17.00 B) \$17.50 C) \$18.00 D) \$18.50	6. C
7. From 21 to 49 is the same as from $(21 - 20)$ to $(49 - 20)$. There are 29. A) 28 B) 29 C) 30 D) 31	7. B
8. $11\times(1+2+3+4+5+6) - 11\times11 = 11\times21 - 11\times11 = 11\times(21 - 11) = 11\times10$. A) 6 B) 7 C) 10 D) 20	8. C
9. Since 20 is one less than 3×7, 20 days is one day less than 3 weeks. A) Sunday B) Monday C) Tuesday D) Wednesday	9. B
10. If I have 4 ten-dollar bills and 1 twenty-dollar bill, my bills are worth \$40 + \$20 = \$60. The average worth is \$60 ÷ 5 = \$12 per bill. A) \$12 B) \$14 C) \$15 D) \$30	10. A
11. Since Brian's neighbor has 3 times as many Hawaiian shirts as he has, the difference between the number of shirts each has is 2 times as many shirts as Brian has. Hence, Brian has $48 \div 2 = 24$ shirts and his neighbor has $24 + 48 = 72$ shirts. A) 24 B) 60 C) 64 D) 72	11. D
12. The number of hours in 10 days is $10 \times 24 = 240$. This is the same as the number of minutes in $240 \div 60 = 4$ hours. A) 4 B) 12 C) 24 D) 30	12. A

Go on to the next page ➠ **5**

	Answers
13. Apollo was born at 1:35 P.M., 50 minutes after Rocky was born. Counting backwards, 35 minutes before 1:35 P.M. is 1 P.M. Another 15 minutes before 1 P.M. is 12:45 P.M. Hence, Rocky was born at 12:45 P.M. A) 12:45 P.M. B) 12:55 P.M. C) 1:25 P.M. D) 2:25 P.M.	13. A
14. 876 rounded to the nearest 10 is 880. 880 multiplied by 2 is 1760. 1760 rounded to the nearest 100 is 1800. A) 1700 B) 1740 C) 1800 D) 1860	14. C
15. The multiples of 9 between 10 and 100 are 2×9, 3×9, . . . , and 11×9. A) 9 B) 10 C) 11 D) 12	15. B
16. The sum of the ten-thousands digit and the hundreds digit of 123456 is 2 + 4 = 6. A) 5 B) 6 C) 7 D) 8	16. B
17. Try each answer. If there are 32 students in all, then 16 wore jeans, 12 wore dresses, and the remaining 4 wore khaki pants. Since 4 out of 32 is one-eighth, there are 32 students in my class. A) 24 B) 26 C) 28 D) 32	\`17. D
18. My family can build 3 snowmen in 45 minutes or 1 snowman in 15 minutes. It would take us 15 × 7 = 105 minutes to build 7 snowmen. A) 1 hr, 5 min B) 1 hr, 15 min C) 1 hr, 25 min D) 1 hr, 45 min	18. D
19. 1×100×3×100×5×100 = 100×(1×3×100×5×100) = 100×(300×500). A) 3 × 5 B) 8 × 100 C) 200 × 15 D) 300 × 500	19. D
20. The 4th song had 30 more notes than the 1st song. Similarly, the 5th song had 30 more notes than the 2nd song and the 6th song had 30 more notes than the 3rd song. If the 1st 3 songs had a total of 2025 notes, the next 3 songs had a total of 2025 + 90 = 2115 notes. A) 2055 B) 2085 C) 2115 D) 3025	20. C
21. The least common multiple of 4 and 14 is 28 and the greatest common factor of 4 and 14 is 2. The difference between 28 and 2 is 26. A) 26 B) 18 C) 10 D) 4	21. A

Go on to the next page ⟫➡ 5

Solution	Answers
22. Of every 8 balloons that Helios sold yesterday, 5 were blue and 3 were white. He might have sold $7 \times 8 = 56$ balloons in all. A) 33 B) 45 C) 56 D) 71	22. C
23. If each citizen speaks 2 languages, the 3000 citizens speak 6000 (not all different) languages. On average, $6000 \div 6 = 1000$ people speak each language. A) 900 B) 1000 C) 1800 D) 2100	23. B
24. Since each square has area 4, the length of a side of each square is 2. The rectangle is 2 by 6, so its perimeter is $2 + 6 + 2 + 6 = 16$. A) 10 B) 12 C) 14 D) 16	24. D
25. The 3 prime numbers between 40 and 50 are 41, 43, and 47. A) 10 and 20 B) 20 and 30 C) 30 and 40 D) 40 and 50	25. D
26. The sum of the lengths of 2 sides of a triangle > the length of the 3rd side. Lengths of 7, 14, and 7 are impossible since $7 + 7$ is not > 14. A) 7 and 8 B) 8 and 9 C) 7 and 14 D) 11 and 11	26. C
27. Since the product of $6 \times 5 \times 4 \times 3 = 360$, Dawn must write 2 non-zero digits. A) 1 B) 2 C) 3 D) 4	27. B
28. Winkle's age in months is 12 times his age in years. Together, the sum is 13 times his age in years. Therefore, Winkle is $2015 \div 13 = 155$ years old. A) 155 B) 160 C) 165 D) 167	28. A
29. If the ones digit is 2, the 2 numbers are 202 and 112. For each ones digit, the number of possible numbers is the same as the ones digit. In all, there are $1 + 2 + 3 + \ldots + 8 + 9 = 45$ numbers. A) 10 B) 25 C) 30 D) 45	29. D
30. It takes 144 workers 60 hours to paint a bridge. That's $144 \times 60 = 8640$ worker-hours. For 108 workers, the job takes $8640 \div 108 = 80$ hours. A) 45 B) 65 C) 80 D) 96	30. C

The end of the contest 5

Visit our Web site at http://www.mathleague.com

Math League Press, P.O. Box 17, Tenafly, New Jersey 07670-0017

Information & Solutions

Spring, 2016

5

Contest Information

- **Solutions** Turn the page for detailed contest solutions (written in the question boxes) and letter answers (written in the *Answer Column* to the right of each question).

- **Scores** Please remember that *this is a contest, and not a test*—there is no "passing" or "failing" score. Few students score as high as 24 points (80% correct); students with half that, 12 points, *deserve commendation!*

- **Answers and Rating Scales** Turn to page 147 for the letter answers to each question and the rating scale for this contest.

Answers

1. The values of the choices are 36, 4, 320, and 1.25, respectively.

   A) $20 + 16$ B) $20 - 16$ C) 20×16 D) $20 \div 16$

   1. C

2. If I eat 3 meals each day, then I eat 21 meals in one week and 42 meals in two weeks.

   A) 6 B) 14 C) 21 D) 42

   2. D

3. $(111 \div 3) \times 3 = 111$.

   A) 37 B) 74 C) 111 D) 333

   3. C

4. Choice A is *not* a whole number.

   A) $0.3 \times 5 = 1.5$ B) $0.4 \times 5 = 2$
   C) $0.5 \times 4 = 2$ D) $3 \times 5 = 15$

   4. A

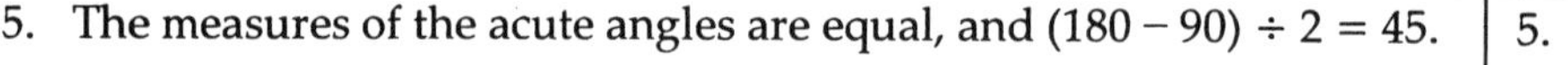

5. The measures of the acute angles are equal, and $(180 - 90) \div 2 = 45$.

   A) 30° B) 45° C) 60° D) 75°

   5. B

6. Six algebra books cost $\$12.50 \times 6 = \$75$. Five geometry books cost $\$14 \times 5 = \$70$. All together, they cost a total of \$145.

   A) \$145.00 B) \$146.50 C) \$150.00 D) \$151.50

   6. A

7. From 31 to 69, there are a total of $69 - 31 + 1 = 39$ whole numbers.

   A) 38 B) 39 C) 40 D) 41

   7. B

8. $25 \times 1 + 25 \times 3 + 25 \times 5 + 25 \times 7 + 25 \times 9 = 25 \times \underline{(1 + 3 + 5 + 7 + 9)}$.

   A) 24 B) 25 C) 125 D) 945

   8. B

9. This is $(1 + 999) + (2 + 998) + (3 + 997) + (4 + 996) + (5 + 995) = 5000$.

   A) 5000 B) 5001 C) 5002 D) 5003

   9. A

10. Since Larry scored 3 times as many points as Mo, Larry's score must be a multiple of 3. Of the choices listed, only choice C is such a multiple.

    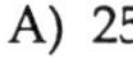

    A) 25 B) 50 C) 75 D) 100

    10. C

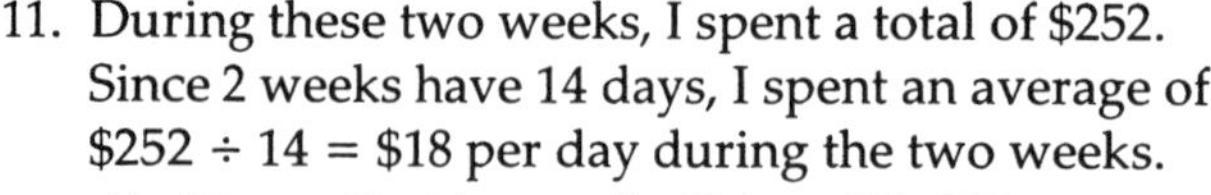

11. During these two weeks, I spent a total of \$252. Since 2 weeks have 14 days, I spent an average of $\$252 \div 14 = \$18$ per day during the two weeks.

    A) 18 B) 36 C) 126 D) 252

    11. A

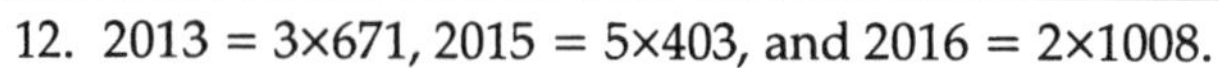

12. $2013 = 3 \times 671$, $2015 = 5 \times 403$, and $2016 = 2 \times 1008$.

    A) 2013 B) 2015 C) 2016 D) 2017

    12. D

Go on to the next page ⟫⟫➡ 5

Answers

13. For every 3 balloons Jack blew up, Jill blew up 5. If Jack blew up $3 \times 7 = 21$ balloons, then Jill blew up $5 \times 7 = 35$ balloons.

A) 23 B) 25 C) 35 D) 56

Answer 13. C

14. According to the restrictions of the question, you can pick any number greater than 0. Try 5: $[(5 + 5^2) \div 5] - 5 = 1$.

A) 0 B) 1 C) 2 D) 3

Answer 14. B

15. If a whole number is divisible by 6, then it must be divisible by 3.

A) 0 B) 1 C) 2 D) 3

Answer 15. A

16. The perimeter of a triangle is the sum of the lengths of its three sides. The length of the third side is $4033 - 2015 - 2016 = 2$.

A) 2 B) 2017 C) 2018 D) 2019

Answer 16. A

17. Tom the Cat catches a mouse every 204 minutes. Today he caught his first mouse at 5:00 A.M. He will catch his third mouse 408 minutes later, or 6 hours and 48 minutes later.

A) 11:24 A.M. B) 11:48 A.M. C) 12:40 P.M. D) 3:12 P.M.

Answer 17. B

18. $3 \times 6 \times 9 \times 12 = 1 \times 2 \times 3 \times 4 \times \underline{3^4} = 1 \times 2 \times 3 \times 4 \times \underline{81}$.

A) 3 B) 9 C) 27 D) 81

Answer 18. D

19. Of the following choices, only 37 is a factor of 111.

A) 5 B) 7 C) 11 D) 37

Answer 19. D

20. The base and the height are 9 and 40, or vice versa. The area of the triangle is $(9 \times 40) \div 2 = 180$.

A) 180 B) 184.5 C) 360 D) 820

Answer 20. A

21. The number of shirts that Tim bought must be a multiple of 12. Of all the choices, only 120 is divisible by 12.

A) 30 B) 40 C) 50 D) 120

Answer 21. D

22. The area of a circle whose radius is 1 is π. The area of a circle whose radius is 2 is 4π.

A) 1 B) 2 C) 4 D) 8

Answer 22. C

Go on to the next page ⟫➡ 5

23. When 50 is divided by 9, the quotient and the remainder are each 5. After writing the phrase 5 times, Ms. Jackson will stop at the 5th letter, I.

A) H B) I C) N D) U

23. B

24. Since 1 m = 100 cm, the area of the square is 100^2 cm^2 = 10^4 cm^2.

A) 10^2 B) 10^4 C) 10^6 D) 10^9

24. B

25. If 12 carrots are left after Jin eats 2/3 of today's carrots, then 12 carrots are 1/3 of the carrots she started with today. So Jin began today with 36 carrots. Since Jin ate 1/2 yesterday, she started with 2 × 36 carrots.

A) 36 B) 48 C) 60 D) 72

25. D

26. 333 unicycles and 666 bicycles have a total of 333 + 2 × 666 = 1665 wheels. The tricycles have a total of 2016 − 1665 = 351 wheels. There are 351 ÷ 3 = 117 tricycles.

A) 117 B) 339 C) 351 D) 999

26. A

27. The sum of even numbers is even, and the sum of an even and an odd is odd, so there can be 1 odd number and 2015 even numbers.

A) 0 B) 1 C) 2014 D) 2015

27. D

28. 1 × 2 × 3 × . . . × 35 × 36 is divisible by 2 × 31 = 62.

A) 37 B) 53 C) 62 D) 82

28. C

29. The numbers are 118, 124, 142, 181, 214, 222, 241, 412, 421, and 811. There are 10 numbers.

A) 8 B) 9 C) 10 D) 11

29. C

30. As the two numbers get closer together, their product gets larger. Since 10 × 40 = 400 and 15 × 35 = 525, the smaller number must be between 10 and 15. Try 11 × 39, 12 × 38, and finally 13 × 37. Since 13 × 37 = 481, choice D is correct.

A) 478 B) 479 C) 480 D) 481

30. D

The end of the contest 5

Visit our Web site at http://www.mathleague.com

6th Grade Solutions

2011-2012 through 2015-2016

Math League Press, P.O. Box 17, Tenafly, New Jersey 07670-0017

Information & Solutions

Tuesday, February 21 or 28, 2012

6

Contest Information

- **Solutions** Turn the page for detailed contest solutions (written in the question boxes) and letter answers (written in the *Answer Column* to the right of each question).

- **Scores** Please remember that *this is a contest, and not a test*—there is no "passing" or "failing" score. Few students score as high as 28 points (80% correct); students with half that, 14 points, *deserve commendation!*

- **Answers and Rating Scales** Turn to page 148 for the letter answers to each question and the rating scale for this contest.

	Answers
1. Terry played an odd number of notes. Odd numbers have a ones digit of 1, 3, 5, 7, or 9, so 2211 is an odd number. A) 2012 B) 2211 C) 3456 D) 4664	1. B
2. The average is 10, so the sum is 20; 20−8 = 12. A) 6 B) 9 C) 12 D) 18	2. C
3. The tens digit of 345 is 4 and the hundreds digit of 456 is 4. Their sum is 8. A) 8 B) 9 C) 10 D) 11	3. A
4. My vacation starts on May 10 and ends on May 20. Subtract 9 from each—it's the same number of days as from May 1 through May 11. A) 9 B) 10 C) 11 D) 12	4. C
5. 2 days = 48 hours = (60 × 48) minutes = 2880 minutes. A) 1440 B) 1660 C) 2000 D) 2880	5. D
6. Multiples of 45 are 45, 90, 135, 180, . . . ; 180 is also a multiple of 12. A) 57 B) 90 C) 180 D) 540	6. C
7. 72 erasers = 72÷12 = 6 dozen erasers; 6 dozen cost Dina 6×50¢ = $3. A) $1.50 B) $3.00 C) $30.00 D) $36.00	7. B
8. The only even prime number is 2, so the odd prime number must be 7. A) 1 B) 5 C) 6 D) 7	8. D
9. The sum of the largest and smallest whole-number divisors of 36 is 36+1. A) 12 B) 15 C) 20 D) 37	9. D
10. If 160 of 400 drivers are late, then 160÷400 = 0.40 = 40% are late for work. A) 20% B) 40% C) 60% D) 80%	10. B
11. Since 220 is divisible by both 4 and 5, there are 220/4 = 55 multiples of 4 between 1 and 222, and 220/5 = 44 multiples of 5. Finally, 55 – 44 = 11. A) 11 B) 22 C) 33 D) 44	11. A
12. 1000 mins = 16 hrs 40 mins. The time is 2:40 A.M. A) 4:40 P.M. B) 10:00 P.M. C) 2:40 A.M. D) 10:00 A.M.	12. C
13. Gil flies 4400 km at 800 km/hour. It will take him (4400÷800) hours = 5.5 hours to complete his trip. A) 4 B) 4.5 C) 5 D) 5.5	13. D

Go on to the next page ⟫⟫➡ 6

14. Work backwards and use inverse operations: 30/2 = 15, 15 + 10 = 25, 25 × 2 = 50, and 50 – 10 = 40. Larry's real age is 40. — 14. B

A) 50 B) 40 C) 30 D) 20

15. A rectangle has a perimeter of 48. If its width is 6 and its length 18, then its perimeter is 6 + 6 + 18 + 18 = 48. — 15. A

A) 6 B) 8 C) 12 D) 16

16. Since 50 quarters = $12.50, and 50 pennies + 50 nickels + 50 dimes = $0.50 + $2.50 + $5.00 = $8.00, I owe her $12.50 – $8.00 = $4.50. — 16. A

A) $4.50 B) $5.50 C) $6.50 D) $7.50

17. 8.5 × 7500 = 63 750. — 17. C

A) 63 000 B) 63 450 C) 63 750 D) 64 005

18. Since there are 3 boys for every 4 girls, the total number of students must be a multiple of 3 + 4 = 7, so the answer is 70. — 18. B

A) 120 B) 70 C) 40 D) 30

19. If the average is $10, the sum is 5×$10 = $50. Without the fifth friend, the friends had $4 all together. The fifth friend has $50 – $4 = $46. — 19. C

A) $9 B) $19 C) $46 D) $49

20. The largest divisor of 180 that is the square of an integer is 36. The smallest divisor of 180 that is the square of an integer is 1; 36 – 1 = 35. — 20. D

A) 5 B) 27 C) 32 D) 35

21. Let's try 14 (or any other number that's 3 more than a multiple of 11). When I divide 14 by 11, the remainder is 3. When I divide 4 × 14 = 56 by 11, the remainder is 1. — 21. A

A) 1 B) 3 C) 7 D) 12

22. $99 \times 88 \times 77 \times 66 \times 55 \div (9 \times 8 \times 7 \times 6 \times 5) = 11^5$. — 22. C

A) 11 B) 11^4 C) 11^5 D) 11^6

23. Since 3:5 = 12:20, the largest whole number for which 3:5 < 12:_?_ is true is 19. — 23. A

A) 19 B) 20 C) 21 D) 24

24. 99 ÷ 9 = 11, so the middle number is 11. Numbers are 7, 8, 9, . . . , 14, 15. — 24. D

A) 9 B) 11 C) 13 D) 15

25. If my house is between my sister's house and school, and if all three are on a straight line, I could be 5 km from school, but no closer than that. — 25. A

A) 4 km B) 5 km C) 8 km D) 10 km

Go on to the next page ⟹ 6

26. Each chemical can be paired with 6 other chemicals for a total of $6 \times 7 = 42$ pairs. However, each pair has been counted twice. So divide by 2: $42 \div 2 = 21$.

    A) 14 B) 21 C) 42 D) 49

    26. B

27. The sum of the digits is 100, which is not divisible by 3 or any multiple of 3.

    A) 4 B) 6 C) 8 D) 20

    27. B

28. Multiply the three smallest primes: $2 \times 3 \times 5 = 30$, whose 8 divisors are 1, 2, 3, 5, 6, 10, 15, and 30.

    A) 3 B) 5 C) 6 D) 8

    28. D

29. $(2^{2000} + 2^{2000}) + (3^{3000} + 3^{3000} + 3^{3000}) = 2\times2^{2000} + 3\times3^{3000} = 2^{2001} + 3^{3001}$.

    A) $4^{2000} + 9^{3000}$ B) $2^{4000} + 3^{9000}$ C) $4^{2001} + 9^{3002}$ D) $2^{2001} + 3^{3001}$

    29. D

30. I drop my nickels at 0 m, 5 m, 10 m, . . . , 90 m, 95 m, and 100 m. In all, I drop 21 nickels. Finally, the value of all 21 of these nickels is $21 \times 5¢ = \$1.05$.

    A) \$0.95 B) \$1.00 C) \$1.05 D) \$1.10

    30. C

31. Since $31^2 = 961 < 1000 < 32^2 = 1024$, the squares of 1, 2, 3, . . . , 31 are less than 1000. Of these 31 squares, 16 are odd and 15 are even.

    A) 15 B) 16 C) 30 D) 31

    31. A

32. One-half of one-quarter is one-eighth. The area of one-eighth of a square is 8; so the total area is 64, the side-length is 8, and the perimeter is 32.

    A) 16 B) 32 C) 48 D) 64

    32. B

33. $(4+8+12+\ldots+400)-(3+6+9+\ldots+300) = (4-3)+(8-6)+(12-9)+\ldots+(400-300) = 1+2+3+\ldots+100 = (1+100)+(2+99)+\ldots+(50+51) = 50\times101 = 5050$.

    A) 100 B) 400 C) 1200 D) 5050

    33. D

34. Write ratios equivalent to 5:3 until a trade yields an 8:7 ratio; 5:3 = 10:6 = 15:9 = 20:12. With 20 clubs and 12 skins, a trade of 4 clubs for 2 skins leaves 16 clubs and 14 skins, which is a 16:14 = 8:7 ratio.

    A) 8 B) 9 C) 10 D) 12

    34. A

35. The 1st 9 remainders are 1, 2, 3, . . . , 8, 0. They repeat 100 times until reaching 999. That's a total of 100×36. Add 1 for 1000.

    A) 3600 B) 3601 C) 4500 D) 4501

    35. B

The end of the contest  **6**

Visit our Web site at http://www.mathleague.com

Math League Press, P.O. Box 17, Tenafly, New Jersey 07670-0017

Information & Solutions

Tuesday, February 19 or 26, 2013

6

Contest Information

- **Solutions** Turn the page for detailed contest solutions (written in the question boxes) and letter answers (written in the *Answer Column* to the right of each question).

- **Scores** Please remember that *this is a contest, and not a test*—there is no "passing" or "failing" score. Few students score as high as 28 points (80% correct); students with half that, 14 points, *deserve commendation!*

- **Answers and Rating Scales** Turn to page 149 for the letter answers to each question and the rating scale for this contest.

	Answers
1. Pete the pilot flew 28 times last month. If 21 of his flights were at night, then 28 − 21 = 7 flights were not at night. A) 7 B) 21 C) 28 D) 49	1. A
2. The sum 12 + 34 + 56 equals each of the following *except* choice D. A) 46+56 B) 12+90 C) 34+68 D) 46+68	2. D
3. If I double the number of pens in my backpack and add 5, I get 23. Subtract 5 and divide by 2 to get (23 − 5) ÷ 2 = 9. A) 9 B) 14 C) 36 D) 56	3. A
4. Distribute subtraction over addition: 65 − (43 + 21) = (65 − 43) − 21. A) 1 B) 12 C) 21 D) 34	4. C
5. One dime and quarter are worth 35¢. One dime less than $1 is 90¢. Since 90¢ − 35¢ = 55¢, the coins in my pocket are worth 55¢. A) 45¢ B) 55¢ C) 65¢ D) 75¢	5. B
6. Five days before Wednesday is Friday. A) Friday B) Sunday C) Monday D) Tuesday	6. A
7. Since each choice is odd, 2 must be one of the addends. A) 11 = 2 + 9 B) 17 = 2 + 15 C) 23 = 2 + 21 D) 31 = 2 + 29	7. D
8. Each of my shoes weighs the same. If 2 of my shoes weigh 12 kg together, then the total weight of 12 of my shoes is 6 × 12 kg = 72 kg. A) 2 kg B) 24 kg C) 36 kg D) 72 kg	8. D
9. 25 × 25 = 5 × 5 × 25. A) 2 B) 5 C) 10 D) 25	9. D
10. (6 × 12) + (12 × 2) = 96 = 32 × 3. A) 48 B) 32 C) 24 D) 12	10. B
11. Since 31 divided by 4 has a remainder of 3, Giggles the Clown could have a total of 31 dots on his costume. A) 31 B) 32 C) 33 D) 34	11. A
12. 420 minutes = 7 hrs.; 7 hrs. before 4 P.M. is 9 A.M. A) 4:00 A.M. B) 7:00 A.M. C) 9:00 A.M. D) 11:40 A.M.	12. C
13. (10 x 100) + (10 x 10) + 10 = 1110. A) 111 B) 1101 C) 1110 D) 101010	13. C

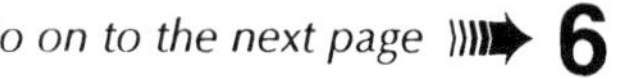
Go on to the next page ➡ 6

Solution	Answers
14. Professor Quack had 7 more students this year than he had last year. Subtract 7 from each choice and then add the result to that choice to see if you get 43: (25 − 7) + 25 = 43. A) 18 B) 25 C) 32 D) 36	14. B
15. In all, 27 trapezoids have 4×27 sides = 108 sides = 3×36 sides, the same number as in 36 triangles. A) 16 B) 18 C) 27 D) 36	15. D
16. There are 6 roses for every 5 daisies in my garden, so 6/(6 + 5) = 6/11 of the 66 flowers I have are roses. Thus, 6/11 × 66 = 36 are roses. A) 11 B) 22 C) 30 D) 36	16. D
17. The sum of two different odd numbers and an even number must be even. A) 52 B) 61 C) 65 D) 77	17. A
18. On a Sunday I put two rabbits in a cage. If the number of rabbits in the cage doubled every day, then I had 4 rabbits, 8 rabbits, 16 rabbits, 32 rabbits, 64 rabbits, 128 rabbits, ... A) Thursday B) Friday C) Saturday D) Sunday	18. C
19. A pomegranate costs as much as 4 pawpaws. If 1 pomegranate costs 50¢ more than 2 pawpaws, then 2 pawpaws cost 50¢ and 4 cost $1. A) 50¢ B) 75¢ C) $1 D) $1.50	19. C
20. Work backwards: 6 × 18 = 108; 108 ÷ 3 = 36. A) 9 B) 36 C) 72 D) 108	20. B
21. The given sum = 11+(12+10)+(13+20)+(14+30)+(15+40)+(16+50) − 150. A) 50 B) 100 C) 150 D) 200	21. C
22. Add 15 to each choice, divide by 3, and add 3 jumps. If the result is the same as the choice, then it's correct. Since (12 + 15) ÷3 + 3 = 12, choice A is correct. A) 12 B) 18 C) 21 D) 24	22. A
23. The value of 10 nickels and 9 dimes is $1.40. The value of 5 quarters is $1.25, and $1.40 − $1.25 =15¢. A) 4 B) 5 C) 14 D) 15	23. D
24. Any odd multiple of 5 has a ones digit of 5. The numbers are 5, 15, 25, . . . , 85, 95. There are 10. A) 9 B) 10 C) 11 D) 19	24. B
25. The remainders are 3, 4, and 1; their sum is 8. A) 3 B) 6 C) 8 D) 12	25. C

Go on to the next page ⟫⟫➡ 6

Solution	Answers
26. There was sunny weather on 12 of 30 days last month; then on 18 days the weather was not sunny. Since 18 ÷ 30 = 0.6, that's 60%. A) 36% B) 40% C) 60% D) 64%	26. C
27. Since \$24 ÷ \$0.80 = 30 and \$24 ÷ \$1.20 = 20, I bought 50 magnets for \$48. Thus, the average cost per magnet was \$48 ÷ 50 = \$0.96. A) \$0.92 B) \$0.96 C) \$1.00 D) \$1.04	27. B
28. The average of 1.75 and 7.25 is equidistant from them. The average is (1.75 + 7.25) ÷ 2 = 4.5. A) 2.75 B) 3.25 C) 3.75 D) 4.5	28. D
29. $2^3 \times 3^4 \times 4^5 \times 6^7 \times 9^{10} = 2^3 \times 3^4 \times 2^{10} \times (2^7 \times 3^7) \times 3^{20} = 2^{3+10+7} \times 3^{4+7+20}$. A) $2^{15} \times 3^{21}$ B) $2^{20} \times 3^{31}$ C) $2^{15} \times 3^{40}$ D) $2^{105} \times 3^{280}$	29. B
30. The ratio of red cars to black cars is 8:5 = 24:15; the ratio of black cars to white cars is 3:4 = 15:20. The minimum number of cars is 24 + 15 + 20 = 59. A) 20 B) 59 C) 74 D) 91	30. B
31. The sum is 25 + 26 + … + 30 = 165. Since 165 ÷ 10 = 16.5, the middle numbers are 16 and 17. The sum is 12 + 13 + … + 16 + 17 + . . . + 20 + 21. A) 17 B) 18 C) 21 D) 26	31. C
32. A radius of a circle with area 36π cm² is 6 cm. The width of the rectangle is 6 cm. A diameter of the circle is 12 cm, so the length of the rectangle is 24 cm. The perimeter of the rectangle is 2 × (6 + 24) = 60 cm. A) 60 cm B) 90 cm C) 144 cm D) 172 cm	32. A
33. For every 3 numbers left, one multiple of 4 was removed. Since 2345 ÷ 3 = 781 R2, 781 multiples of 4 were removed. Since there is a remainder of 2, the last number in the list was 4 × 781 + 2 = 3126. A) 3126 B) 3127 C) 3129 D) 3130	33. A
34. Each day I loaded 90 boxes instead of 120, I was 30 boxes short. If I were on schedule, I would need to load 720 boxes the last 6 days. I had to load 480 extra boxes. Since 480 ÷ 30 = 16, I had 16 + 6 = 22 days to finish this temporary job. A) 10 B) 16 C) 22 D) 26	34. C
35. Working backwards, I counted 2/3 the number of leaves on each previous day. So on Sunday, I counted $(2/3)^5 \times 2430$ = 320 leaves. A) 160 B) 240 C) 280 D) 320	35. D

The end of the contest 6

Visit our Web site at http://www.mathleague.com

Steven R. Conrad, Daniel Flegler, and *Adam Raichel,* contest authors

SIXTH GRADE MATHEMATICS CONTEST

Math League Press, P.O. Box 17, Tenafly, New Jersey 07670-0017

Information & Solutions

Tuesday, February 18 or 25, 2014

6

Contest Information

- **Solutions** Turn the page for detailed contest solutions (written in the question boxes) and letter answers (written in the *Answer Column* to the right of each question).

- **Scores** Please remember that *this is a contest, and not a test*—there is no "passing" or "failing" score. Few students score as high as 28 points (80% correct); students with half that, 14 points, *deserve commendation!*

- **Answers and Rating Scales** Turn to page 150 for the letter answers to each question and the rating scale for this contest.

2013-2014 6TH GRADE CONTEST SOLUTIONS

	Answers
1. The band's trombone plays 2013 notes, the trumpet plays 2014 notes, and the tuba plays 218 notes. That is a total of 2013 + 2014 + 218 = 4245 notes. A) 6245 B) 6045 C) 4245 D) 645	1. C
2. 30 + 30 + 30 = 90 = 3 × 30. A) 10 B) 30 C) 90 D) 270	2. B
3. This has the same remainder as 1 divided by 3. The remainder is 1. A) 0 B) 1 C) 2 D) 3	3. B
4. One dollar = 20 nickels, one quarter = 5 nickels, one dime = 2 nickels, and one nickel = 1 nickel. This is the same as 28 nickels. A) 140 B) 70 C) 35 D) 28	4. D
5. 20 − 5 × 2 = 20 − 10 = 10 = 2 × 5. A) 5 B) 15 C) 25 D) 30	5. A
6. Three weeks after yesterday, which was a Monday, is a Monday also. Two days after a Monday is a Wednesday. A) Monday B) Tuesday C) Wednesday D) Thursday	6. C
7. (2 × 2) × (2 × 3) × (2 × 4) × (2 × 5) = 2 × 3 × 4 × 5 × (2 × 2 × 2 × 2). A) 2 B) 6 C) 8 D) 16	7. D
8. The number 25.12 is closer to 30 than it is to 20. A) 18.26 B) 21.42 C) 24.68 D) 25.12	8. D
9. When I split the cost of a video game equally with 4 friends, we each pay \$12. It costs the five of us 5 × \$12 = \$60. If only 4 of us split the cost, we each pay \$60 ÷ 4 = \$15. We each pay \$15 − \$12 = \$3 more. A) \$3.00 B) \$4.00 C) \$15.00 D) \$16.00	9. A
10. The number halfway between 4 and 26 is (4 + 26) ÷ 2 = 15. A) 11 B) 13 C) 15 D) 17	10. C
11. At 5:00 P.M. on Friday, Hal got locked in. Since 5040 mins. is 5040 ÷ 60 = 84 hrs., Hal got out in 3 days 12 hrs. That's Tuesday at 5 A.M. A) Sunday B) Monday C) Tuesday D) Wednesday	11. C
12. 1 000 000 = 10 × 10 × 10 × 10 × 10 × 10 = 10^6. A) 10^5 B) 10^6 C) 10^7 D) 10^8	12. B
13. 1.5 × 20 = 30 = 2 × 15 = 200% of 15. A) 15 B) 25 C) 30 D) 40	13. A

Go on to the next page ⟹ 6

14. The cannon fired its cannonball a distance of 2000 mm. Since 1 cm = 10 mm, 2000 mm is the same as $2000 \div 10 =$ 200 cm.

A) 20000 B) 200 C) 20 D) 2

14. B

15. The area of four 2×8 rectangles is $4 \times 16 = 64 =$ area of one 8×8 square.

A) 8 B) 16 C) 32 D) 64

15. C

16. Since $231 = 3 \times 7 \times 11$, the sum of its prime factors is $3 + 7 + 11 = 21$.

A) 21 B) 22 C) 152 D) 383

16. A

17. The markers numbered 1 through 19 and the markers numbered 81 through 100 are all more than 30 km from the one numbered 50. That is a total of $19 + 20 = 39$ markers.

A) 38 B) 39 C) 40 D) 41

17. B

18. $5 \times 3 = 15$.

A) 8 B) 10 C) 12 D) 15

18. D

19. Of 60 people, 24 are men and 36 are women. The ratio of men to women at the meeting is 36:24 = 3:2.

A) 3:2 B) 2:3 C) 11:6 D) 6:11

19. A

20. The sum of the positive divisors of 18 is $1 + 2 + 3 + 6 + 9 + 18 = 39$.

A) 18 B) 20 C) 38 D) 39

20. D

21. I rode my bicycle a total of 450 km in 15 hours. At this rate I would travel 30 km in 1 hour and 15 km in 30 minutes.

A) 15 B) 30 C) 60 D) 900

21. A

22. In 8 years Mac's grandmother will be twice as old as she was 29 years ago. So 29 years ago she was $29 + 8 = 37$. She is now 66; in 4 years, she'll be 70.

A) 62 B) 66 C) 70 D) 74

22. C

23. If the sum of 7 consecutive integers is 63, the middle one is $63 \div 7 = 9$. The desired sum is $6 + 12 = 18$.

A) 18 B) 24 C) 42 D) 64

23. A

24. $\sqrt{3^2 + 4^2} = \sqrt{9 + 16} = \sqrt{25} = 5$.

A) 5^2 B) 3×4 C) $3 + 4$ D) 5

24. D

25. Of the first 100 positive whole numbers, 12 are multiples of 8, and 25 are multiples of 4. The ratio is 12:25.

A) 2:1 B) 12:25 C) 13:25 D) 1:2

25. B

Go on to the next page ⟹ **6**

	Answers
26. Craig the Croc cried at 15% of the movies he saw, so he didn't cry at 85% of them. Since 85% of the movies is 34, 100% of the movies is $34 \div 0.85 = 40$, and 15% is 6. A) 6 B) 12 C) 18 D) 40	26. A
27. Consecutive number pairs with a different # of digits must be like (9,10), (99,100), ... Their product is divisible by $3 \times 5 = 15$. A) 4 B) 11 C) 15 D) 100	27. C
28. The bookshelf contains $18\,000 \div 360 = 50$ books. There are $50 \div 5 =$ 10 books on each shelf. A) 10 B) 50 C) 55 D) 250	28. A
29. $2^3 \times 2^5 \times 2^7 \times 2^{11} = 2^{3+5+7+11} = 2^{26} = (2^2)^{13} = 4^{13}$. A) 4^{13} B) 16^{26} C) 2^{1155} D) 16^{1155}	29. A
30. The 7 vertical stripes of red, orange, yellow, green, blue, indigo, and violet are repeated 50 times. Since $100 \div 7 = 14$ R2, there are 14 full repetitions of this pattern in 100 stripes and then one more red and one more orange. A) 7 B) 12 C) 14 D) 15	30. D
31. $(4) \times (4) + (6) \times (3) = 16 + 18 = 34$. A) 24 B) 28 C) 34 D) 66	31. C
32. At 6:15, hr. hand has moved $(1/4) \times 30°$ from 6, so $\angle = 90° + 30°/4 = 97.5°$. A) 82.5° B) 90° C) 97.5° D) 270°	32. C
33. Let the herbal teas be *A*, *B*, *C*, *D*, *E*, and *F*. *A* can be combined with 9 teas, *B* with 8 besides *A*, *C* with 7 besides *A* and *B*, . . . , and *F* with 4 besides *A*, *B*, *C*, *D*, and *E*. The total # of blends is $9 + 8 + \ldots + 4 = 39$. A) 24 B) 27 C) 39 D) 54	33. C
34. It takes 36 workers 48 hours to paint a ship. In all, that is $36 \times 48 = 1728$ worker-hours. With 24 workers, it would take $1728 \div 24 = 72$ hours to paint the ship. A) 32 B) 56 C) 64 D) 72	34. D
35. The product of all whole numbers from 1 through 35 includes only 5 factors of 7. A) 2^{18} B) $7^6 \times 3$ C) $3^{15} \times 5$ D) $10^8 \times 7$	35. B

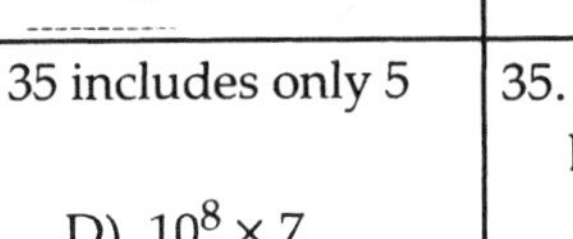

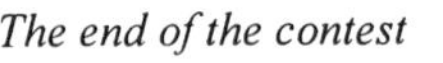

The end of the contest

6

Visit our Web site at http://www.mathleague.com

Math League Press, P.O. Box 17, Tenafly, New Jersey 07670-0017

Information & Solutions

Tuesday, February 17 or 24, 2015

6

Contest Information

- **Solutions** Turn the page for detailed contest solutions (written in the question boxes) and letter answers (written in the *Answer Column* to the right of each question).

- **Scores** Please remember that *this is a contest, and not a test*—there is no "passing" or "failing" score. Few students score as high as 28 points (80% correct); students with half that, 14 points, *deserve commendation!*

- **Answers and Rating Scales** Turn to page 151 for the letter answers to each question and the rating scale for this contest.

Solution	Answers
1. Three cases of 24 cans each = 3×24 cans = 72 cans = 12×6 cans = twelve boxes of 6 cans each. A) 6 B) 15 C) 21 D) 96	1. A
2. A trapezoid has 4 sides. A) 3 B) 4 C) 5 D) 10	2. B
3. Abby put an X through 7 of the 28 days in this month. Abby put an X through $7 \div 28 = 0.25$ days. A) 0.07 B) 0.25 C) 0.28 D) 0.35	3. B
4. $60 \div 4 = 15 = 3 \times 5$. A) 20 B) 12 C) 6 D) 5	4. D
5. $30 \times 40 \times 50 = (3 \times 10) \times 40 \times 50 = (3 \times 40) \times 10 \times 50 = 120 \times 500$. A) 50 B) 200 C) 500 D) 600	5. C
6. There are $24 \div 4 = 6$ groups of 4 students. In each of these groups, 3 students have brown eyes. In all, 6×3 students have brown eyes. A) 6 B) 8 C) 18 D) 21	6. C
7. The side-length of each small square is $36 \div 4 = 9$. The side-length of the large square is 18. Its perimeter is $4 \times 18 = 72$. A) 36 B) 48 C) 72 D) 144	7. C
8. $80 + (160 + 240) \div 4 = 80 + (40 + 60) = 40 + 80 + (120 \div 2)$. A) 4 B) 2 C) 1 D) 0	8. B
9. The remainders are (in order) 7, 3, 3, and 4. A) $1111 \div 8$ B) $2222 \div 7$ C) $3333 \div 6$ D) $4444 \div 5$	9. A
10. Since $14 = 7 \times 2$, 7 is a factor of the product. A) 13 B) 11 C) 9 D) 7	10. D
11. Thok spends 12 hours of the day in the cave. Of the remaining 12 hours, he spends 3 hours on the hunt. That leaves $12 - 3 = 9$ hours remaining to watch films. A) 3 B) 6 C) 9 D) 25	11. C
12. $(2 \times 3) \times 6 \times 6^2 \times (2 \times 3) \times 6 \times 6^2 = 6^8$. A) 6^5 B) 6^6 C) 6^7 D) 6^8	12. D
13. The total value of my 15 coins is 5¢ + 20¢ + 75¢ + \$1 + \$1 = \$3.00. The average value of one of my coins is \$3.00 ÷ 15 = \$0.20. A) \$0.20 B) \$0.60 C) \$1.50 D) \$3.00	13. A

14. Wyatt weighs as much as 2 hats. His sheep weighs as much as 4 hats. In all, Wyatt, his sheep, and his hat weigh as much as 7 hats. If 7 hats weigh 210 kg, 1 hat weighs 30 kg and 2 hats (same as Wyatt) weigh 60 kg.

A) 30 kg B) 35 kg C) 60 kg D) 70 kg

14. C

15. The lcm of 24 and 30 is 120. The gcf of 24 and 30 is 6. Their difference is 114.

A) 114 B) 117 C) 234 D) 237

15. A

16. Of two consecutive whole numbers, one is odd and the other even. Their sum is odd.

A) a prime B) a perfect square
C) a multiple of 13 D) an even number

16. D

17. The greatest prime factor of $(2 \times 2 \times 3 \times 5) \times (2 \times 2 \times 2 \times 3) \times 7$ is 7.

A) 5 B) 7 C) 12 D) 21

17. B

18. By the distributive prop., $(12+34)\times(56+78) = 12\times(56+78) + 34\times(56+78)$.

A) 12 B) 34 C) 56 D) 78

18. B

19. Seth eats an apple at 4 P.M. on Monday. Since $100 \div 24 = 4R4$, it's 4 days and 4 hours later or 8 P.M. on Friday when he eats another one.

A) noon B) 4 P.M. C) 8 P.M. D) midnight

19. C

20. If 2 flocks = 5 flecks, then 500 flocks = 250 × 2 flocks = 250 × 5 flecks.

A) 200 B) 250 C) 1000 D) 1250

20. D

21. $\left(\sqrt{64}+\sqrt{64}\right)^2 = (8+8)^2 = (16)^2 = 256$.

A) 16 B) 64 C) 128 D) 256

21. D

22. Since 70 + 60 + 50 = 180, the largest angle in the triangle is 70°.

A) 70° B) 75° C) 80° D) 90°

22. A

23. Rob rode at an average rate of 30 km/h = 30000 m/60 min = 500 m/min.

A) 30 B) 105 C) 120 D) 500

23. D

24. Since 65:100 = 13:20, Clara will lay a total of 20 eggs. She has laid 13 eggs; 7 eggs are left to be laid.

A) 6 B) 7 C) 13 D) 20

24. B

25. The average value (which in this case is also the middle number) is $182 \div 7 = 26$. The three preceding numbers are 24, 22, and 20.

A) 20 B) 23 C) 26 D) 32

25. A

26. Subtract by a large multiple of 8 to quickly count down by 8s to be near choices given: $777 - 640 = 137$. Subtract 8 again: $137 - 8 = 129$.

A) 123 B) 125 C) 127 D) 129

26. D

27. Five apples cost $5 \times 15¢$ more than 5 pears, so 1 pear costs 75¢. The cost of 5 apples and 6 pears = the cost of 12 pears, so the cost is $12 \times 75¢ = \$9$.

A) \$3 B) \$6 C) \$9 D) \$18

27. C

28. The desired difference is $(1 \times 3 \times 9 \times 27) - 27 = (27 \times 27) - (1 \times 27) = (27 - 1) \times 27 = 26 \times 27$.

A) 2 B) 27 C) 2×27 D) 26×27

28. D

29. Since $2^6 < 100 < 2^7$, the prime factorization of a whole number less than 100 is the product of at most 6 primes.

A) 3 B) 4 C) 5 D) 6

29. D

30. If the shaded rectangle is 8 by 1, then the square is 8 by 8 and the entire figure is 8 by 9. The area of the entire figure is then $8 \times 9 = 72$. No greater area is possible.

A) 24 B) 64 C) 72 D) 81

30. C

31. The 1st letter Gabriel wrote was G, and every 7th letter after was G. The 99th letter he wrote was G, and the 100th was an "a."

A) a B) b C) r D) i

31. A

32. At the end of year 1, I had \$110. At the end of year 2, I had \$121. At the end of year 3, I had \$133.10. At the end of year 4, I had \$146.41. At the end of year 5, I had (to the nearest dollar) \$161.

A) \$162 B) \$161 C) \$160 D) \$150

32. B

33. Mrs. Andrews' 200 6-kg bags contain a total of 1200 kg of seed. One-fourth of this is sunflower seed, so $1200 \text{ kg} \div 4 = 300$ kg is sunflower seed. If sunflower seed comes in 12-kg bags, she needs $300 \div 12 = 25$ such bags.

A) 25 B) 34 C) 50 D) 67

33. A

34. The product of 100 100s is 100^{100}. This is the same as $100^{99} \times 100^1$ or the sum of 100^{99} 100s.

A) 100^{100} B) 100^{99} C) 100^{10} D) 100^2

34. B

35. $5^{21} \times 4^{11} \div 2 = 5^{21} \times 2^{22} \div 2^1 = 5^{21} \times 2^{21} = (5 \times 2)^{21} = 10^{21}$.

A) 10^{11} B) 10^{21} C) 10^{22} D) 10^{23}

35. B

The end of the contest  **6**

Visit our Web site at http://www.mathleague.com

Math League Press, P.O. Box 17, Tenafly, New Jersey 07670-0017

Information & Solutions

Tuesday, February 16 or 23, 2016

6

Contest Information

- **Solutions** Turn the page for detailed contest solutions (written in the question boxes) and letter answers (written in the *Answer Column* to the right of each question).

- **Scores** Please remember that *this is a contest, and not a test*—there is no "passing" or "failing" score. Few students score as high as 28 points (80% correct); students with half that, 14 points, *deserve commendation!*

- **Answers and Rating Scales** Turn to page 152 for the letter answers to each question and the rating scale for this contest.

	Answers
1. $2 \times 2016 = 2 \times (4 \times 504) = (2 \times 4) \times 504 = 8 \times 504$. A) 504 B) 508 C) 1008 D) 8064	1. A
2. Bert Sampson sold his 2 beavers for a total of \$444.44. Mho gave him \$500 for them. The change is \$500.00 — \$444.44 = \$55.56. A) \$277.78 B) \$277.88 C) \$55.56 D) \$55.66	2. C
3. The sum of the measures of the angles in a trapezoid is always 360°. A) 90° B) 180° C) 270° D) 360°	3. D
4. $10 \times 20 \times 30 \times 40 = (1 \times 2 \times 3 \times 4) \times 10^4 = 24 \times 10^4$. A) 10^3 B) 10^4 C) 10^5 D) 10^6	4. B
5. This is $(1 + 999) + (2 + 998) + (3 + 997) + (4 + 996) = 4000$. A) 3998 B) 3999 C) 4000 D) 4001	5. C
6. $2016 = 7 \times 288 = 8 \times 252 = 9 \times 224$. A) 10 B) 9 C) 8 D) 7	6. A
7. The tenths digit is 6 and the hundredths digit is 7. Their sum is 13. A) 7 B) 11 C) 13 D) 15	7. C
8. 5 out of 6 of the 2016 lightbulbs are not defective. Thus $2016 \times 5/6 =$ 1680 lightbulbs are not defective. A) 5 B) 336 C) 1680 D) 2016	8. C
9. Divide $12012 \div 24$ to get remainder 12. 12 hours after 7 A.M. is 7 P.M. A) 1 A.M. B) 1 P.M. C) 7 A.M. D) 7 P.M.	9. D
10. For any 2 such primes, the factors are 1, the primes, and their product. A) 3 B) 4 C) 5 D) 6	10. B
11. Adam ate 5/20 of the pancakes, Jerry ate 7/20, and Steve ate 6/20, for a total of 18/20. That means Dan ate (20 — 18)/20 = 2/20 of them. A) Adam B) Jerry C) Steve D) Dan	11. B
12. The greatest common factors of the 4 pairs of numbers are 2, 5, 3, and 1, respectively. A) 4 & 18 B) 5 & 25 C) 6 & 33 D) 8 & 35	12. D

Go on to the next page ⟫➡ **6**

13. I donate a total of \$100 + 2 × \$50 + 3 × \$20 + 4 × \$10 + 5 × \$5 = \$325. Each person receives \$325 ÷ 5 = \$65.

    A) \$37 B) \$65 C) \$70 D) \$75

    13. B

14. Paul begins 24 m behind Peter. The distance between them decreases by 3 m per second. Paul needs 24÷3 = 8 seconds to catch Peter.

    A) 6 B) 7 C) 8 D) 14

    14. C

15. The product of two different nonzero integers can never be 0.

    A) prime B) zero C) even D) composite

    15. B

16. The largest two-digit factor is $3^2 \times 11 = 99$.

    A) 66 B) 77 C) 88 D) 99

    16. D

17. The number of hours in 10 days is 240; 240 minutes is 240÷60 = 4 hours.

    A) 2 B) 3 C) 4 D) 6

    17. C

18. The sum of any multiple of 100 and 84 is divisible by 4.

    A) 4 B) 8 C) 9 D) 16

    18. A

19. Each pair has 2 intersection points. The 6 pairs have 12 such points.

    A) 16 B) 12 C) 8 D) 6

    19. B

20. 50/20 = 2.5 = 2.5 × 100% = 250%.

    A) 40 B) 140 C) 200 D) 250

    20. D

21. The sum of any even number of odd integers is always even.

    A) 2016 B) 2015 C) 1 D) 0

    21. A

22. Divide 180 into 1 + 3 + 6 = 10 equal parts. One angle measures 18°, and the other angles measure 3 × 18° = 54° and 6 × 18° = 108°.

    A) 108° B) 72° C) 54° D) 18°

    22. A

23. Suppose the side-length is 1. (Choose any length.) The perimeter is 4 and its square is 16. The area of the square is 1. The quotient is 16 ÷ 1 = 16.

    A) 2 B) 4 C) 8 D) 16

    23. D

24. 1000 m/s = 1 km/s = 3600 km/hr.

    A) 60 B) 360 C) 3600 D) 6000

    24. C

25. Since $C = 2\pi r$ and $r = \pi$, $C = 2\pi \times \pi = 2\pi^2$.

    A) π B) 2π C) π^2 D) $2\pi^2$

    25. D

Go on to the next page ➡ **6**

26. Clark got 1/3 = 3/9 of the coins. Lois got 1/3 of 2/3 = 2/9 of the coins. That is 3/9 + 2/9 = 5/9, so Lex had 4/9 left. Since 4/9 of the coins is 144, 2/9, of the coins is 72. A) 72 B) 48 C) 36 D) 32	26. A
27. The ones digit of a 4th power can be 0, 1, 5, or 6. It can never be 3. A) 1 B) 3 C) 5 D) 6	27. B
28. Continue subtracting 4¢ from 2458¢ until the difference is a multiple of 9. This happens when the difference is 2430¢ which is 270 × 9¢. A) 273 B) 272 C) 271 D) 270	28. D
29. I scored a total of 720 on all 8 tests. The total of 435 on the first 5 tests leaves a total of 285 for the last 3 tests, so the average is 285 ÷ 3 = 95. A) 96 B) 95 C) 94 D) 93	29. B
30. The cost of the calculus book is \$21. The cost of my geometry book is \$21 × (4/3) = \$28. My algebra book costs \$28 × 2 = \$56. All 3 books cost \$21 + \$28 + \$56 = \$105 in total. A) \$28 B) \$56 C) \$84 D) \$105	30. D
31. A multiple of 3, 4, and 5 is a multiple of 60, and $16\times60 < 1000 < 17\times60$. A) 14 B) 15 C) 16 D) 17	31. C
32. Each number in the second sum is 18 greater than the corresponding number in the first sum. Thus the second sum is 1935 + 18 × 9 = 2097. A) 2015 B) 2017 C) 2097 D) 2099	32. C
33. $3^{336} \times 9^{336} \times 27^{336} = 3^{336} \times 3^{672} \times 3^{1008} = 3^{336+672+1008} = 3^{2016}$. A) 3^{1008} B) 3^{1344} C) 3^{1680} D) 3^{2016}	33. D
34. A unit circle's area is π. A circle with $r = 1.5$ has area $2.25\pi = \pi + 1.25\pi$. A) 50 B) 100 C) 125 D) 150	34. C
35. By the Counting Principle, there are a total of 26 × 26 = 676 possible initials. If each possibility were assigned to 2 people, there would be 676 × 2 = 1352 people. The next person would be the 3rd person to share a possibility. A) 1353 B) 1352 C) 677 D) 676	35. A

The end of the contest **6**

Visit our Web site at http://www.mathleague.com

Answer Keys & Difficulty Ratings

2011-2012 through 2015-2016

ANSWERS, 2011-12 4th Grade Contest

1. D	7. A	13. C	19. A	25. D
2. B	8. C	14. D	20. C	26. B
3. B	9. D	15. B	21. B	27. B
4. B	10. B	16. A	22. A	28. D
5. D	11. C	17. C	23. D	29. C
6. A	12. A	18. C	24. C	30. A

RATE YOURSELF!!!

for the 2011-12 4th GRADE CONTEST

Score	Rating
28-30	Another Einstein
25-27	Mathematical Wizard
23-24	School Champion
19-22	Grade Level Champion
17-18	Best In The Class
15-16	Excellent Student
12-14	Good Student
9-11	Average Student
0-8	Better Luck Next Time

ANSWERS, 2012-13 4th Grade Contest

1. A	7. D	13. D	19. C	25. D
2. B	8. B	14. A	20. C	26. A
3. D	9. B	15. C	21. B	27. B
4. D	10. C	16. B	22. A	28. D
5. C	11. C	17. A	23. A	29. C
6. C	12. D	18. B	24. B	30. A

RATE YOURSELF!!!

for the 2012-13 4th GRADE CONTEST

Score	Rating
29-30	Another Einstein
26-28	Mathematical Wizard
24-25	School Champion
21-23	Grade Level Champion
19-20	Best In The Class
16-18	Excellent Student
14-15	Good Student
12-13	Average Student
0-11	Better Luck Next Time

ANSWERS, 2013-14 4th Grade Contest

1. A	7. B	13. C	19. D	25. B
2. D	8. C	14. B	20. B	26. A
3. B	9. C	15. C	21. C	27. C
4. A	10. A	16. B	22. C	28. B
5. D	11. B	17. B	23. D	29. D
6. D	12. C	18. A	24. A	30. C

RATE YOURSELF!!!

for the 2013-14 4th GRADE CONTEST

Score	Rating
29-30	Another Einstein
27-28	Mathematical Wizard
24-26	School Champion
22-23	Grade Level Champion
20-21	Best In The Class
17-19	Excellent Student
15-16	Good Student
13-14	Average Student
0-12	Better Luck Next Time

HOT STUFF

ANSWERS, 2014-15 4th Grade Contest

1. A	7. B	13. B	19. C	25. B
2. D	8. B	14. A	20. B	26. A
3. D	9. A	15. D	21. C	27. D
4. A	10. C	16. A	22. D	28. B
5. B	11. D	17. C	23. A	29. D
6. C	12. C	18. A	24. B	30. C

RATE YOURSELF!!!

for the 2014-15 4th GRADE CONTEST

Score	Rating
29-30	Another Einstein
26-28	Mathematical Wizard
23-25	School Champion
21-22	Grade Level Champion
18-20	Best In The Class
16-17	Excellent Student
13-15	Good Student
11-12	Average Student
0-10	Better Luck Next Time

ANSWERS, 2015-16 4th Grade Contest

1. B	7. A	13. C	19. C	25. A
2. D	8. C	14. A	20. B	26. C
3. D	9. B	15. A	21. A	27. D
4. C	10. C	16. B	22. A	28. C
5. B	11. C	17. C	23. C	29. B
6. B	12. B	18. D	24. B	30. A

RATE YOURSELF!!!

for the 2015-16 4th GRADE CONTEST

Score	Rating
27-30	Another Einstein
24-26	Mathematical Wizard
21-23	School Champion
18-20	Grade Level Champion
15-17	Best In The Class
13-14	Excellent Student
11-12	Good Student
9-10	Average Student
0-8	Better Luck Next Time

ANSWERS, 2011-12 5th Grade Contest

1. D	7. A	13. D	19. A	25. C
2. B	8. B	14. A	20. A	26. B
3. C	9. C	15. B	21. C	27. A
4. B	10. B	16. C	22. C	28. B
5. D	11. B	17. A	23. A	29. D
6. C	12. D	18. C	24. C	30. A

RATE YOURSELF!!!

for the 2011-12 5th GRADE CONTEST

Score	Rating
28-30	Another Einstein
26-27	Mathematical Wizard
23-25	School Champion
21-22	Grade Level Champion
19-20	Best In The Class
17-18	Excellent Student
14-16	Good Student
12-13	Average Student
0-11	Better Luck Next Time

ANSWERS, 2012-13 5th Grade Contest

1. A	7. C	13. D	19. B	25. B
2. A	8. C	14. A	20. C	26. A
3. C	9. C	15. D	21. C	27. A
4. B	10. B	16. D	22. B	28. D
5. D	11. B	17. B	23. D	29. D
6. A	12. A	18. C	24. B	30. D

RATE YOURSELF!!!

for the 2012-13 5th GRADE CONTEST

Score	Rating
28-30	Another Einstein
26-27	Mathematical Wizard
24-25	School Champion
22-24	Grade Level Champion
20-21	Best In The Class
18-19	Excellent Student
15-16	Good Student
13-14	Average Student
0-12	Better Luck Next Time

ANSWERS, 2013-14 5th Grade Contest

1. D	7. A	13. C	19. B	25. C
2. C	8. D	14. B	20. B	26. D
3. B	9. B	15. B	21. D	27. C
4. A	10. C	16. D	22. B	28. A
5. B	11. D	17. A	23. A	29. D
6. A	12. C	18. C	24. D	30. B

RATE YOURSELF!!!

for the 2013-14 5th GRADE CONTEST

Score	Rating
29-30	Another Einstein
27-28	Mathematical Wizard
25-26	School Champion
22-24	Grade Level Champion
20-21	Best In The Class
18-19	Excellent Student
16-17	Good Student
14-15	Average Student
0-13	Better Luck Next Time

ANSWERS, 2014-15 5th Grade Contest

1. C	7. B	13. A	19. D	25. D
2. A	8. C	14. C	20. C	26. C
3. B	9. B	15. B	21. A	27. B
4. D	10. A	16. B	22. C	28. A
5. A	11. D	17. D	23. B	29. D
6. C	12. A	18. D	24. D	30. C

RATE YOURSELF!!!

for the 2014-15 5th GRADE CONTEST

Score	Rating
29-30	Another Einstein
27-28	Mathematical Wizard
25-26	School Champion
22-24	Grade Level Champion
20-21	Best In The Class
17-19	Excellent Student
15-16	Good Student
12-14	Average Student
0-11	Better Luck Next Time

ANSWERS, 2015-16 5th Grade Contest

1. C	7. B	13. C	19. D	25. D
2. D	8. B	14. B	20. A	26. A
3. C	9. A	15. A	21. D	27. D
4. A	10. C	16. A	22. C	28. C
5. B	11. A	17. B	23. B	29. C
6. A	12. D	18. D	24. B	30. D

RATE YOURSELF!!!

for the 2015-16 5th GRADE CONTEST

Score	Rating
28-30	Another Einstein
26-27	Mathematical Wizard
23-25	School Champion
20-22	Grade Level Champion
18-19	Best In The Class
15-17	Excellent Student
13-14	Good Student
11-12	Average Student
0-10	Better Luck Next Time

ANSWERS, 2011-12 6th Grade Conteste

1. B	8. D	15. A	22. C	29. D
2. C	9. D	16. A	23. A	30. C
3. A	10. B	17. C	24. D	31. A
4. C	11. A	18. B	25. A	32. B
5. D	12. C	19. C	26. B	33. D
6. C	13. D	20. D	27. B	34. A
7. B	14. B	21. A	28. D	35. B

RATE YOURSELF!!!

for the 2011-12 6th GRADE CONTEST

Score	Rating
33-35	Another Einstein
30-32	Mathematical Wizard
26-29	School Champion
23-25	Grade Level Champion
20-22	Best In The Class
17-19	Excellent Student
15-16	Good Student
12-14	Average Student
0-11	Better Luck Next Time

ANSWERS, 2012-13 6th Grade Contest

1. A	8. D	15. D	22. A	29. B
2. D	9. D	16. D	23. D	30. B
3. A	10. B	17. A	24. B	31. C
4. C	11. A	18. C	25. C	32. A
5. B	12. C	19. C	26. C	33. A
6. A	13. C	20. B	27. B	34. C
7. D	14. B	21. C	28. D	35. D

RATE YOURSELF!!!

for the 2012-13 6th GRADE CONTEST

Score	Rating
34-35	Another Einstein
32-33	Mathematical Wizard
29-31	School Champion
26-28	Grade Level Champion
23-25	Best In The Class
20-22	Excellent Student
17-19	Good Student
14-16	Average Student
0-13	Better Luck Next Time

ANSWERS, 2013-14 6th Grade Contest

1. C	8. D	15. C	22. C	29. A
2. B	9. A	16. A	23. A	30. D
3. B	10. C	17. B	24. D	31. C
4. D	11. C	18. D	25. B	32. C
5. A	12. B	19. A	26. A	33. C
6. C	13. A	20. D	27. C	34. D
7. D	14. B	21. A	28. A	35. B

RATE YOURSELF!!!

for the 2013-14 6th GRADE CONTEST

Score	Rating
33-35	Another Einstein
30-32	Mathematical Wizard
27-29	School Champion
24-26	Grade Level Champion
21-23	Best In The Class
18-20	Excellent Student
15-17	Good Student
12-14	Average Student
0-11	Better Luck Next Time

ANSWERS, 2014-15 6th Grade Contest

1. A	8. B	15. A	22. A	29. D
2. B	9. A	16. D	23. D	30. C
3. B	10. D	17. B	24. B	31. A
4. D	11. C	18. B	25. A	32. B
5. C	12. D	19. C	26. D	33. A
6. C	13. A	20. D	27. C	34. B
7. C	14. C	21. D	28. D	35. B

RATE YOURSELF!!!

for the 2014-15 6th GRADE CONTEST

Score	Rating
34-35	Another Einstein
31-33	Mathematical Wizard
28-30	School Champion
25-27	Grade Level Champion
22-24	Best In The Class
19-21	Excellent Student
16-18	Good Student
12-15	Average Student
0-11	Better Luck Next Time

ANSWERS, 2015-16 6th Grade Contest

1. A	8. C	15. B	22. A	29. B
2. C	9. D	16. D	23. D	30. D
3. D	10. B	17. C	24. C	31. C
4. B	11. B	18. A	25. D	32. C
5. C	12. D	19. B	26. A	33. D
6. A	13. B	20. D	27. B	34. C
7. C	14. C	21. A	28. D	35. A

RATE YOURSELF!!!

for the 2015-16 6th GRADE CONTEST

Score	Rating
33-35	Another Einstein
29-32	Mathematical Wizard
26-28	School Champion
23-25	Grade Level Champion
21-22	Best In The Class
19-20	Excellent Student
17-18	Good Student
15-16	Average Student
0-14	Better Luck Next Time

WE MADE IT!